DAS

Get **more** out of libraries

Please return or renew this item by the last date shown.

You can renew online at www.hants.gov.uk/library

Or by phoning 0845 603 5631

Hampshire
County Council

"Nicely written, thoroughly researched, and highly recommended.
Doomsday is already marked in the calendar."

Alberto Fairen – NASA Ames Research Center

ABOUT THE AUTHORS

Dr David Darling is an astronomer, freelance science writer, and creator of one of the most popular online encyclopedias of space and astrobiology. He is the author of the bestselling *Equations of Eternity*.

Dr Dirk Schulze-Makuch is Professor in Astrobiology at Washington State University. His research has been widely published in media ranging from academic journals to *The New Scientist*.

Together they are the authors of the critically-acclaimed *We Are Not Alone*.

MEGACATASTROPHES!
Nine Strange Ways the World Could End

DAVID DARLING & DIRK SCHULZE-MAKUCH

ONEWORLD

A Oneworld Book

First published by Oneworld Publications in 2012
This paperback edition published in 2012

Copyright © David Darling and Dirk Schulze-Makuch 2012

The moral rights of David Darling and Dirk Schulze-Makuch to
be identified as the Authors of this work has been asserted by them in
accordance with the Copyright, Designs and Patents Act 1988

Paperback ISBN 978-1-185168-947-7
Ebook ISBN 978-1-78074-027-0

Typeset by Jayvee, Trivandrum, India
Cover design by Richard Green
Printed and bound by Nørhaven, Denmark

Oneworld Publications
185 Banbury Road
Oxford OX2 7AR
England

CONTENTS

ILLUSTRATIONS

ACKNOWLEDGEMENTS

Our thanks go to our editor at Oneworld, Mike Harpley, for his insightful suggestions and enthusiasm, Louis Irwin and Ed Guinan for helpful advice on some of the chapters, and Jeff Darling for rendering a number of the illustrations. Most of all, we are grateful to our families for their love and support while this book was being prepared.

INTRODUCTION

We live in a quiet backwater of an ordinary, middle-aged spiral galaxy, circling one of the more sedate kinds of stars, on the surface of a stunningly life-friendly planet. Nature has dealt us a very kind hand indeed. But that doesn't mean we're completely safe or that we can afford to be complacent.

Space rocks as big as houses zip by us, closer than the Moon, every few months or so; some the size of large mountains have smashed into the Earth in the past causing serious mayhem. Giant stars explode, supervolcanoes erupt, ice ages come and go, the very fabric of space and time might rip apart at any moment if some theories are to be believed. And if these potential natural disasters aren't enough, there are threats of our own making in the form of new technologies that could spin out of control. We might even fall victim to an alien attack – seriously – just like in the good old B-movies of the 1950s.

Not that there's any need to panic. Almost certainly the world won't end this week, or in our lifetime, or in our children's children's lifetimes. But it's possible, even if we manage to avoid pressing the nuclear button, that the human race, and even the Earth itself, could be wiped out in less time than it takes you to read these pages. Or the end might come more gradually but still uncomfortably fast, over the space of a few years or a few centuries. The destruction might involve more than just our planet, too. There are ways in which the entire universe, with its hundreds of billions of galaxies scattered across billions of light years, could slip swiftly into oblivion without warning – and with more than a little human help.

• • • •

Prophesying the end of the world has always been a popular pastime. At the time of writing, the more excitable Internet forums, the tabloid press, and late-night radio talk-shows are gripped by tales of a coming apocalypse. December 2012 is the fateful month, some say, when, according to an ancient calendar of the Mayans, everything is going to go horribly pear-shaped. But before blowing your life's savings on one last outrageous street party, it may be worthwhile reading the small print on this prediction.

The calendar in question is what's called the Mesoamerican Long Count calendar, which supposedly began 5,125 years ago. In 1966, the American archaeologist and anthropologist Michael D. Coe argued that the current, fourth age of the Long Count would come to an end on December 21, 2012, and that this date would have been highly portentous to the Mayans living centuries ago. "There is a suggestion," he wrote, "that Armageddon would overtake the degenerate peoples of the world and all creation on the final day ... [O]ur present universe [would] be annihilated when the Great Cycle of the Long Count reaches completion."

Although Coe remains a respected figure, most Mayan specialists today disagree with him on this particular issue. The consensus view among scholars is that, although the Mayans believed in world ages, they made no predictions about any catastrophe happening at the end of the present cycle. Nothing in their records suggests anything other than that they saw it as the prelude to yet another world age, which, in their minds, would have been a cause for celebration rather than fear.

Pseudoscientists and pseudohistorians, however, aren't prone to letting the opinions of experts get in the way of a good story. The 2012 Phenomenon is alive and kicking, and will doubtless remain so until December 22 when we wake up to find that the Earth hasn't been knocked off its axis, or suffered any other ill effects from obscure cosmic alignments, geomagnetic reversals, or close flybys of unknown planets. If past experience is anything to go by, explanations of how we survived will quickly emerge, and new doomsday forecasts will appear to replace the ones that didn't quite work out. And, as always, these revised dire warnings will find a ready and eager audience.

Apocalyptic predictions have been doing the rounds for thousands of years. An Assyrian clay tablet dating back to about 2800 BCE warned: "Our earth is degenerate in these latter days. There are signs that the world is speedily coming to an end." Doom-mongers throughout the ages have had a uniformly dismal success rate.

Often, as in the case of the 2012 prophesy, the date when everything will unravel is pinpointed with great precision – to the embarrassment of the author when the world continues on its merry way past the appointed time. Early in Rome's history, many Romans feared the city would be destroyed in the 120th year of its founding. There was a myth that a dozen eagles had revealed to Romulus a mystical number representing the lifetime of Rome, and it came to be supposed that each eagle represented ten years. Since the Roman calendar began in what is our 753 BCE, the date of destruction was put at 634 BCE.

In more recent times, hardly a year has gone by without someone staking their (often dubious) reputation on a prediction that the Day of Judgment or the "Rapture" is nigh. The year 2010, for instance, was the last time anyone needed to print calendars according to the Hermetic Order of the Golden Dawn. In 2011 it was the turn of Harold Camping and his California-based Family Radio group to declare that May 21 would see the Second Coming of Christ. Camping's followers emptied their bank accounts trying to spread the word and were less than pleased to find themselves still stuck on a very material world and totally broke. Meanwhile New Hampshire atheist Bart Centre made a killing by setting up Eternal Earth-Bound Pets and charging clients $135 a time for insurance policies that guaranteed care for animals whose owners had ascended to a better place. British physicist Brian Cox tweeted: "I think we should all pretend the rapture is happening so that when Harold Camping gets left behind later today he'll be livid."

Astronomical events have long been favored as portents of doom. Comets especially have been singled out as celestial omens. The 1066 return of Halley's Comet in the skies over Britain (recorded in the Bayeux Tapestry) certainly didn't coincide with good luck for King Harold II of England, who took an arrow in the eye, but may have been viewed in a more kindly light by his adversary William of Normandy. The same object was still rattling people's nerves several centuries later when the Italian scholar Platina wrote, "A hairy and fiery star having made its appearance for several days, the mathematicians declared that there would follow grievous pestilence, dearth, and some great calamity."

The appearance of another comet – Kohoutek – stirred David Berg (a.k.a. Moses David) of the Children of God cult to red-ink January 1974 as the fateful month. Unfortunately for his credibility, Kohoutek failed to collide with the Earth, although that didn't stop him making further predictions about the Second Coming.

John Gribbin and co-author Stephen Plagemann would probably like to forget the year 1982, because it was then that, according to their book *The Jupiter Effect*, the planets would align in such a way that giant earthquakes, tsunami, and other unpleasantness would be visited upon us. In fact, nothing untoward happened – there was never any good astronomical reason it should – and a suitably chastened Gribbin went on to write many more popular science books (including the slightly less successful *The Jupiter Effect Reconsidered*), though none based on quite such sensational claims.

On July 19, 1993, Sister Marie Gabriel Paprocski announced her prophecy that a comet would collide with Jupiter on or before July 25, 1994, causing the "biggest cosmic explosion in the history of mankind" and bringing an end to the world. In fact, a comet *did* hit Jupiter on July 16, 1994; however, Sister Marie's prediction seems a little less startling in view of the fact that it came nearly two months after astronomer Brian Marsden reported that comet Shoemaker-Levy 9 was on track for an impact with the giant planet. And, of course, there's no way that a relatively little object breaking up in Jupiter's atmosphere could have any effect at all on us hundreds of millions of kilometers away.

Tragedy accompanied the claim by the San Diego-based Heaven's Gate cultists that a UFO trailing behind comet Hale–Bopp would save their souls from the imminent destruction of the world. This rumor started when amateur astronomer Chuck Shramek mistook a star for what he thought was a "Saturn-like object" following the comet; Internet gossip ballooned the story to ludicrous proportions. There was no UFO and Hale–Bopp never came close to colliding with Earth, but thirty-nine members of the cult committed suicide in the belief that their souls would be among the few to be rescued.

It seems we never learn. The year 2011 saw more folk getting agitated about the imminent arrival of comet Elenin. Google "Elenin" and you'd be bombarded with ill-informed offerings about how Elenin would come close to or even ram into the Earth, bringing death and destruction on a biblical scale. In fact, Elenin was nothing special in astronomical terms – just another comet visiting the inner parts of the solar system from the depths of space. With a nucleus, or hard central part, only a few kilometers wide, it was actually on the small side as comets go. And astronomers knew well in advance that the closest it would come to us, on October 16, 2011, was thirty-five million kilometers, or about ninety times further away than the Moon. Shortly after its discovery, a few armchair theorists had mistaken the size of Elenin's coma – the glowing, almost vacuum-thin shiny fuzz of vaporized particles around the nucleus – for the size of the nucleus itself. Some of these self-styled experts then caused a stir on the Internet by claiming that Elenin was as big as a planet and would cause chaos during its close passage of the Earth. As the world now knows, little comet Elenin broke apart and disappeared from view last year without unduly disrupting our affairs.

Nothing captures the imagination quite as much as the terrifying prospect of worlds in collision. The Russian-born American psychiatrist Immanuel Velikovsky achieved remarkable popular success with a series of books on this theme. Scouring ancient literary sources, including the Bible, for his data, Velikovsky pieced together a revisionist chronology of events involving close encounters between the Earth and other planets, notably Venus and Mars. His *Worlds in Collision* (1950) and *Ages in Chaos* (1952) were best-sellers in their day, and such was the intricacy of his claims that it took scientists the best part of thirty years to critique them fairly and in full. Although Velikovsky didn't emerge from the analysis well, it's generally recognized that he uncovered some interesting references in ancient texts to the possible effect of comets and meteorites.

Another author who claimed to have found evidence of past collisions, from Sumerian writings, was the Azeri-born American Zecharia Sitchin. Like the German writer Erich von Däniken (of *Chariots of the Gods* fame), Sitchin was a proponent of the ancient astronaut hypothesis, which maintains that extraterrestrials have influenced human history. His theory was that Sumerian culture wasn't home-grown, as previously assumed by every respectable archaeologist and historian, but was in fact seeded by the Anunnaki, a race of aliens from a world whose existence he announced in his book *The Twelfth Planet* (1976) – Nibiru. No amount of debunking of Sitchin's faux-erudite work by Sumerian scholars, who've pointed out multiple flaws in his translations and his selective reading of texts, diminished the popularity of this entertaining narrative. Sitchin's books have been translated into more than twenty-five languages and have sold millions of copies.

Myth builds upon myth, fed by an insatiable appetite for escapist re-tellings of science and history. We seem to love a good scary story, especially if someone who poses as an authority tells us it might be true.

• • • •

Many apocalyptic predictions come about because of a shaky understanding of astronomy or the laws of physics. But what does real science have to say about the prospect of megacatastrophes?

Firstly, again, there's no need to lose sleep over these things. To the best of our knowledge, Earth isn't due to be demolished to make way for some hyperspatial express route, as it was in *The Hitchhiker's Guide to the Galaxy*. Yet, there's no denying, it does occasionally come under fire from space. As recently as 1908 an extraterrestrial missile, probably a wayward chunk of a comet, vaporized explosively in the atmosphere above Siberia, flattening trees over an area of several hundred square kilometers. Reindeer were the main casualties on that occasion but if the intruder had come down over a major city it would have been a very different story. Asteroids *do* hit our planet. The most infamous of them smashed into what is now Mexico's Yucatan peninsula 65.5 million years ago and caused such environmental mayhem that it played a role (though perhaps not an exclusive one) in finishing off the dinosaurs, along with two-thirds of all other species living at the time. We'll be hit again in the future, inevitably, unless a planetary defense system is put in place. It's just a matter of when and how hard.

More devastating still could be a gamma-ray blast originating far outside the Solar System. Sometimes when big stars explode they give off, in their death throes, an almost unimaginably intense pulse of high-energy radiation that tears across space at the speed of light. No inhabited planet within several hundred light years of such a blast would escape unscathed. A nearby gamma-ray blast could rip away most of Earth's protective ozone shield leaving us at the mercy of DNA-busting solar ultraviolet.

Nothing so dramatic will happen to our own star. The Sun, by cosmic standards, is a respectable, well-behaved stellar citizen. Even so, it does have its moments of unpredictability and, over longer periods, its energy output changes with major consequences for life on Earth. The ins and outs of solar variability are something that we're only just beginning to get a handle on – and what we're learning is a little disturbing. At the very least, an up-tick in solar flares during the next solar cycle, which is just starting, could disrupt sat-nav devices which have become ubiquitous since the last cycle and which rely on being able to pick up incredibly weak satellite radio signals. These signals could be drowned out by an influx of solar radiation, making everything from car navigation to oil-tanker docking a hit-and-miss affair.

Slightly more worryingly, automatic landing systems for aircraft and military missions could be compromised.

In one way or another, our increasing dependence on technology is a big risk factor for modern man. We're building vast instruments, such as the Large Hadron Collider (LHC) – all twenty-seven kilometers of it, buried up to 170 meters below the Franco-Swiss border – to explore realms of physics about which we have little understanding (indeed, that's the whole point). Will the LHC spawn little black holes that will gobble up the Earth? Almost certainly not, since such mini-holes are probably being created all the time by cosmic-ray collisions in the atmosphere with no obvious ill effect. Will it bring to light the much-vaunted Higgs boson? Well, hopefully so, because if it does we'll know we're on the right track with our most promising ideas about the make-up of the subatomic particle zoo. On the other hand, a couple of reputable physicists have already suggested that the universe abhors Higgs bosons so much that it's trying to tell us so through various early problems that the LHC has experienced. In other words, effects are rippling back through time that prevent the Higgs from being created in the first place. Although slightly tongue-in-cheek, their suggestion does highlight how little we know about science under extreme conditions.

That's the trouble with new science and technology: because it's new and we don't properly understand it, it can have unforeseen consequences. Nanotechnology is one of the upcoming developments that, some claim, might have very nasty side effects. Sophisticated, self-replicating machines the size of protein molecules might be a godsend to medical science (enabling miniature disease-zapping submarines to patrol the bloodstream and so forth) but a calamity for civilization as a whole if they go AWOL and start converting the environment into the dreaded "grey goo", like a Borg assimilation of nature at the microscopic level. As it is, some types of nanoparticles have already been earmarked as health hazards because of the way they enable toxic substances to build up in the bodies of lab animals and potentially to damage DNA.

These are just some of the scientifically credible but lesser-known threats that *Homo sapiens* faces at this interesting but risky phase in its history. We'll be looking at each of nine categories of danger, the level of hazard they pose to our planetary ecosystem and humans in particular, and what, if anything, we could do to mitigate them.

You might be wondering why we haven't mentioned more obvious problems on the horizon. Two of the most immediate threats to our survival, on everyone's minds right now, are climate change and terrorism or other forms of global conflict. For the simple reason that these have been written about so much already we'll almost completely ignore them here. They're hugely important, of course, and rightly dominate much of our news. But our aim in this book is to take a rather more light-hearted look at other Doomsday scenarios. These are the lesser-known, and sometimes slightly eccentric, ways that our species might meet a sticky end.

MEET THE CATASTROPHOMETER

To get an idea of how likely and threatening are each of the types of megacatastrophe we'll be talking about, the "Catastrophometer" appears at the end of each chapter. This will show a value from 0 (absolutely no need to worry) to 10 (be very, very afraid).

The Catastrophometer applies only to megacatastrophes that might happen over the next 100 years or so (although the book itself sometimes deals with larger timescales). Generally, the higher the reading the more likely is the event and the bigger its anticipated effect on the human population.

The scale is crude and you may think it odd that a "moderate" risk of 10 million people meeting a sticky end ranks higher than a "low" probability of a total wipe-out – but we wanted to keep it simple. Also, the values shown at the end of each chapter are just the authors' personal opinions and not to be taken too seriously! But hopefully this little device will offer a rough guide to the threat level and, if nothing else, a topic for conversation.

Figure 1 The Catastrophometer. Credit: Patrick Knowles.

The Catastrophometer scale is to be interpreted as follows:

Rating	Probability (of megacatastrophe)	Loss of human life
0	zero	0
1	low	10 million or more
2	low	1 billion or more
3	low	total extinction
4	moderate	10 million or more
5	moderate	1 billion or more
6	moderate	total extinction
7	high	10 million or more
8	high	1 billion or more
9	high	total extinction
10	certain	total extinction

CHAPTER 1

NANO NIGHTMARE

In 1959, two years after the first artificial satellite climbed into orbit, the brilliant, bohemian US physicist Richard Feynman gave a talk at an American Physical Society meeting at Caltech called "There's Plenty of Room at the Bottom." It was entertaining and mind-stretching, as were all Feynman's offerings. More importantly, it marked the intellectual beginnings of a new field that became known as nanotechnology – an endeavor that promises a cornucopia of scientific miracles or, if things go badly wrong, a quick and messy end to the Earth and everything on it.

The prefix "nano" comes from the Greek *nanos* for "dwarf." In metric units it means a billionth; so a nanometer is a billionth of a meter. It's a handy unit for measuring the size of atoms and molecules. For instance, a hydrogen atom – the smallest of all atoms – is about a tenth of a nanometer (0.1 nm) across, while a molecule of hemoglobin (the substance that carries oxygen in our blood) is about 5 nm across. Nanotechnology, a term coined by Tokyo Science University professor Norio Taniguchi in 1974, got its name because it has to do with manipulating matter at the atomic and molecular level, typically in the 1 to 100 nm range.

"Consider," said Feynman in his talk, "the final question as to whether, ultimately – in the great future – we can arrange the atoms the way we want; the very atoms, all the way down!" At the time his suggestion must have seemed more like science fiction than anything we could ever actually achieve, but in the 1980s American engineer K. Eric Drexler began promoting the ideas and potential of nanotechnology with evangelical zeal in speeches and a best-selling book, *Engines of Creation: the Coming Era of Nanotechnology*.[1]

Drexler was inspired early in his career by grand innovative schemes such as space colonies and other ambitious solutions to impending human crises – over-population, resource depletion, and the like. Nanotechnology he saw as a panacea to many of the world's problems. What he called "universal assemblers" (microscopic machines that build atom by atom) could be used, he suggested, to make everything from miniature medical subs that would patrol a patient's blood vessels, delivering drugs or clearing blocked arteries, to molecular-sized environmental scrubbers that took pollutants out of the air.

But even Drexler, extreme advocate of what he saw as one of the next great revolutions, was alive to the perils of meddling at the molecular level. He envisaged the possibility of self-replicating nanomachines – devices that could make perfect copies of themselves – being released into the environment to tackle problems like oil spills or air pollution. If these went awry and began breaking down all kinds of other molecules, they could quickly and disastrously reduce the whole Earth to a ball of uniform nano-sludge he called "grey goo."

MATERIAL MENACE

Nanotechnology, although still in its infancy, is now real and very much upon us in the form of consumer products. One of the earliest branches of it to reach fruition and have a practical impact is that of nanomaterials – substances whose unique properties stem from features, such as the shape of their molecules, on the nanoscale (less than about 100 nm across). The most talked-about nanomaterials are the fullerenes, named after maverick American engineer, architect, and futurist Richard Buckminster Fuller. Many molecules of fullerenes – cage-like arrangements of carbon atoms – resemble the geodesic domes that Fuller invented.

Buckminsterfullerene, the first fullerene to be discovered, has sixty carbon atoms locked into a soccer-ball shape. Other common members of the family are built around 70, 72, 76, 84, or 100 carbon atoms. More recently, another type of fullerene has come to light known as carbon nanotubes or buckytubes, which take the form of cylinders and look, on the molecular level, like incredibly long rolls of chicken wire. In fact, they are made essentially by rolling up sheets of graphite (the stuff in pencil "lead"), just one atom thick. Size for size, nanotubes are by far the strongest and stiffest materials yet found: as tough in terms of tensile strength as a cable one millimeter in diameter that could support a weight of about six-and-a-half metric tons. They're also incredibly light, making them ideal for use in up-market tennis rackets and bicycle frames.

Carbon nanotubes have other exceptional properties that make them seem almost magical by comparison with ordinary materials. For example, one nanotube can be nested precisely within another slightly wider tube and then slid, almost frictionlessly, in and out like a folding telescope or turned around its axis to create a rotational bearing. The latter property has been used as

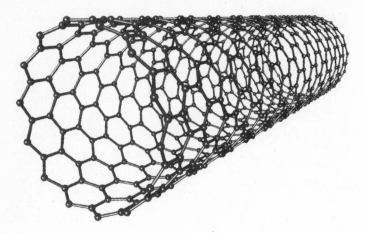

Figure 2 A buckytube – a rolled sheet of hexagonally-linked carbon atoms. Credit: Michael Ströck/Wikipedia.

the basis of the world's smallest spinning motor, which is also one of the first genuine products of molecular nanotechnology – the precise positioning of atoms to create useful machines as originally envisaged by Richard Feynman.

Unfortunately, fullerenes have a less wholesome side to them. From a legal standpoint, buckyballs and buckytubes are currently pigeon-holed in the same category as common-or-garden graphite – one of the familiar forms of carbon (along with diamond and soot). But a number of studies have raised serious questions about the safety of carbon as a nanomaterial. Research published in 2004 by environmental toxicologist Eva Oberdörster, at Southern Methodist University in Texas, pointed to major brain damage in fish that had been exposed to fullerenes for just two days in concentrations as low as 0.5 parts per million, similar to the doses of other pollutants found in coastal bays.[2] The fish were also found to have altered gene markers (sections of DNA whose inheritance can be tracked) in their livers, showing that their entire physiology had been affected. In another test, fullerenes killed water fleas, a significant link in the marine web of life.

Earlier studies, in 2002, ominously revealed that nanoparticles gradually built up in the bodies of lab animals,[3] and still other studies showed that fullerenes can move freely through soil and be taken in by earthworms – a potential link in the food chain to humans.

Nor are the risks confined to carbon nanoparticles. Research at the University of California in San Diego in 2002 demonstrated that cadmium selenide nanoparticles, also called quantum dots, can cause cadmium poisoning in people. In 2004 British scientist Vyvyan Howard reported that gold nanoparticles might be able to travel through a mother's placenta to the fetus;[4] silver is also used in nano form, as an antimicrobial agent in everything from paint and toothpaste to teddy bears, but its effects on the body aren't known. Nanometal oxides are blended into ceramics and coatings, making them more durable, but again without any real knowledge of what physiological damage they might be causing. As far back as 1997 scientists at Oxford found that nanoparticles used in sunscreen created free radicals that disrupt DNA.

Generally, the smaller the nanoparticles, the more toxic and bioactive they seem to be. As their size diminishes, to the level of what's called ultrafine pollution, they're able to interact with living systems more because they can easily slip through the skin or into the lungs, and from there into the bloodstream and other organs. A University of Rochester study found that nanoparticles could even cross the crucial blood-brain barrier and collect in the brain's olfactory bulb (part of the forebrain involved in the perception of smell).[5] Once inside the deepest parts of the body, these minuscule invaders trigger unexpected biochemical reactions that wreak havoc within cells.

Companies are investing billions of dollars each year in products, including make-up, energy drinks, car paint, high-tech clothing, and sports equipment, that contain nanoparticles. Yet only a paltry effort is being put into figuring out the consequences for our health and the wider environment. Our bodies, and those of other animals, have built-in defenses against many potentially harmful substances found in nature. But nanomaterials behave very differently than the same elements and compounds in their normal forms, and nanotechnology is coming up with new and exotic substances so quickly that there's not enough time to research their long-term effects properly.

There's no authority to regulate nanotech-based products. Given the insidious nature of nanoparticles – their talent for slipping past our bodies' defenses – and their ability to accumulate steadily in tissues and even affect DNA, we may literally be storing up problems for the future.

Most at risk, obviously, are workers involved in actually making stuff that contains nanoparticles. Millions of people around the world are already exposed to high doses of nano-dust because of their jobs, and the number is set to rise swiftly. Comparisons have been made to the case of asbestos dust. But whereas it took many years for the carcinogenic properties of asbestos to be recognized, British researchers have already reported that carbon nanotubes, injected into mice, caused the same kind of damage as asbestos.[6] A follow-up study by the US National Institute for Occupational Safety and Health found that when mice inhaled nanotubes, the molecules migrated from their lungs to the surrounding tissue, in the same way that asbestos causes the rare cancer known as mesothelioma.

Possible green tech applications of nanomaterials include cheaper, more efficient solar panels and water-filtration systems, energy-saving batteries, and lighter vehicles that use less fuel. But the fear is what these substances, with their invasive molecules, might do once they leak away into the environment, down drains and into the water and soil. Some of them may even make it into our water supply, and once ingested these nanomolecules have virtually unlimited access to every organ and tissue in our body. If a significant fraction of the human population, or animals in the food chain, ingests a substance that over time proves toxic or, through its effects on genetic material, is harmful to future generations, the long-term effects could be catastrophic.

MINIATURE MARVELS

Waiting in the wings is another and more far-reaching aspect of nanotechnology. This is molecular nanotechnology, or molecular manufacturing (MM) as it's also called – the building of machines and other devices, atom by atom, as Richard Feynman first described over half a century ago. Researchers are still working on the basics of this new field of engineering but it's already been compared in potential to the Industrial Revolution, with the difference that it might take off within years rather than decades.

The scanning tunneling microscope (STM), invented in 1981, was the first instrument to allow matter to be imaged at the level of individual atoms. Eight years later, IBM researcher Don Eigler used an STM to become the first person in history to move and control an individual atom. Within weeks of this achievement, he and his team used a custom-built version of the device to position thirty-five xenon atoms individually so that they spelled the company's initials.

In 2009 a group of physicists at the University of Madrid used a close relative of the STM, called the atomic force microscope (AFM), to build simple structures atom by atom. Unlike the IBM team's method, which relied on pushing or pulling atoms

Figure 3 The letters "IBM" spelled out using 35 xenon atoms. Credit: IBM.

from the surface of a material using the tip of the microscope and which had to be done at very low temperatures, the new technique is based on the controlled swapping of an atom at the instrument's tip for an atom on the surface when the two are near enough. Using the atoms at the tip as a kind of ink, it's possible to write or draw with the microscope. The swapping process can be repeated in different positions over the surface, thus gradually making complicated structures. It marks an important step in realizing Feynman's dream of building extraordinary things, atomically precise, "all the way down".

What Feynman may not have had in mind was an automobile: after all, a car measured in nanometers isn't much use to human drivers. But a nanocar, of sorts, is exactly what a team of chemists at Rice University assembled in 2005.[7] The four wheels of the car, consisting of buckyballs, are held together by an H-shaped chassis of organic molecules, which moves forward and back on its fullerene wheels when heated to 200°C. Far from trying to corner the market for transporting tiny, tiny people, the Rice team simply wanted to know how fullerenes move around on metal surfaces.

A big step forward in molecular manufacturing will be to build actual machines on a nanoscale with motors that can carry out useful tasks. Plenty of progress has already been made in this

Figure 4 A nanocar built by a team of engineers at Rice University. The molecule of which the car is made has an H-shaped chassis with fullerene groups attached at the four corners to act as wheels. Credit:: Anatoly Kolomeisky, Rice University.

direction. The world's smallest steam engine, complete with three pistons driven by steam from a little reservoir of heated water, measures just five microns (5,000 nanometers) across and works just like a full-size one. The engine was built at Sandia Laboratories, in Albuquerque, New Mexico, from the same technology used to make silicon chips and microprocessor components. Chip fabrication methods have also been brought to bear in fashioning microscopic levers, rotors, gears, ratchets, and other components of engines far too small to be seen with the unaided eye.

In a two-pronged attack, nanotechnologists are using methods that work on the smallest of scales (such as atomic force microscopy) and methods that deal with larger amounts of substances (like those applied in making silicon chips) to usher in the age of practical molecular manufacturing. High-speed computing and healthcare may be among the early beneficiaries of this coming revolution, but the fact is that molecular-level engineering will affect almost every human activity and restructure the global economy. A June 2003 report prepared for investors by Credit Suisse First Boston concluded:

> Nanotechnology is a classic, general-purpose technology (GPT). Other GPTs, including steam engines, electricity, and railroads, have been the basis for major economic revolutions. GPTs typically start as fairly crude technologies, with limited uses, but then rapidly spread into new applications. All prior GPTs have led directly to major upheavals in the economy ... And nanotechnology may be larger than any of the other GPTs that preceded it. Because of the advent of nanotechnology ... [t]he majority of the companies in today's Dow Jones industrials Index are unlikely to be there 20 years from now.

Much of the modern world may be on the verge of being rebuilt from the atom up. Nanotechnology will infiltrate our lives to an extent even greater than electricity, telecommunications, and the Internet. It could be wonderful. On the other, it's slightly unnerving to think of something working away at the level of individual atoms and molecules, even inside our own bodies, that we don't properly understand.

ENGINES OF DESTRUCTION

The potential risks of molecular manufacturing are as great as its widely touted benefits. On the one hand, MM will allow speedy prototyping and cheap manufacture of a whole slew of powerful products. On the other, it's likely to arrive on the scene, as a fully fledged practical technology, alarmingly fast once the initial steps have been mastered – too fast for us to adapt to its implications.

Almost inevitably, one of the first sectors to exploit the power of the new field will be the military. With the ability to make surveillance devices, detonators, and some types of weapons microscopically small, nations and armed groups around the world will enter a new and deadly arms race. Nano-devices designed to explode on commercial airliners might prove almost impossible to detect without unacceptably invasive detection methods. Chemical, biological, and radiogenic weapons as small as barely visible insects will be aimed at individuals or other small targets with deadly effect. Consider that a dose of botulism – a bacterial and potentially lethal disease that causes creeping paralysis – sufficient to kill a person weighs a mere 100 nanograms. A hundred times that amount would fit inside the volume of the smallest known insect, and as many as fifty billion toxin-carrying devices – enough to wipe out everyone on the planet several times over – could be stashed inside a holdall.

Nano-weapons will make today's nuclear arms look clumsy, expensive, hard to come by, and inflexible. Addressing the 1995 Foresight Conference on Molecular Nanotechnology, retired Admiral David E. Jeremiah, Vice-Chairman of the US Joint Chiefs of Staff, said: "Military applications of molecular manufacturing have even greater potential than nuclear weapons to radically change the balance of power." Unlike nuclear weapons, MM devices could be delivered unnoticed in advance to where they're intended to go off and be better targeted. What's more, unless their development were tightly controlled, the number of nations and groups able to build and deploy them could mushroom alarmingly, increasing the chance of regional conflicts becoming global in scope.

Once molecular manufacturing becomes a standard tool for making commercial products the environment will be at risk from "nano-litter", composed of a toxic mixture of discarded nano-gadgetry and nanomachine-made waste. Today, giant landfills bear testimony to the ease with which we throw away macroscopic products and other cast-offs, packed with still-useful and highly processed materials. Tomorrow, when nanotechnology can spawn microscopic consumer products and materials, with no limits on choice or application, at virtually zero cost per unit, the temptation to discard with complete abandon will be overwhelming. If past experience is anything to go by, collective environmental damage and associated health risks will be part and parcel of the era of cheap, throw-away nano-manufacturing.

Nanotechnology is proving to be a very different type of science from anything that's come before. It happily tramples across conventional boundaries between disciplines and unashamedly focuses on applications at the expense of fundamental understanding. Given the massive resources being poured into nanotechnology in the Far East, it may also be the first major technology of the modern era to be developed mainly outside the US and Europe. That will make it harder to regulate, at least for those in the western world, and its consequences far less predictable.

GOO-ING DOWN

In the 1960s the Hungarian-born American mathematician John von Neumann wrote about machines that could make exact copies of themselves. He envisaged a kind of robot equipped with a computer brain that could be programmed to reproduce itself from raw materials taken from its surroundings. The copy would be so perfect that it too would carry the instructions needed for making further clones, so that the process could continue indefinitely.

It wasn't long before some people suggested that von Neumann machines, in the form of robot spacecraft, would be an ideal way for an advanced civilization to learn more about the

Galaxy without having to venture out in person. Simply build a von Neumann probe and launch it toward a nearby star. Upon arrival the probe would replicate itself, over and over again, down to the last nut and bolt, from materials found on the surface of one of the star's planets. The new probes would then set off for other stars and their worlds, where they would reproduce again, spawning another generation of self-replicating spacecraft. Given not-unreasonable assumptions about how fast the probes could travel, calculations showed that all the stars in the Galaxy could be visited and explored in this way within a few million years.

But then someone pointed out that if this were such a great way for a smart race to find out about the Milky Way, then we ought to have come across some of these von Neumann probes by now. In fact, given the ease with which they could copy themselves almost ad infinitum, they should be pretty much everywhere. The Solar System ought to be like a junkyard or parking lot of alien self-replicating vessels. Because it obviously isn't, that could be taken as a sign that there aren't any intelligent extraterrestrials anywhere in the Galaxy.[8]

Not so fast, replied astronomer Carl Sagan.[9] Any aliens capable of building star-faring von Neumann probes would be clever enough to realize the danger of launching them in the first place: namely that, in time, in slavish obedience to their programs, the spacecraft would convert almost all the matter in the Galaxy into von Neumann probes!

The information technology that drives our mobile phones or laptops already has nanoscale components. The latest generation of microprocessors, for example, uses transistors that measure only a few tens of nanometers across. Another few decades of development and there'll be true nanocomputers which, working at the heart of tiny assemblers of the type first described by Eric Drexler, may make possible von Neumann machines of microscopic dimensions.

At first sight, such devices look just the job – ideal, for instance, for mopping up after oil spills. A tailor-made von Neumann chemical vacuum cleaner might seem perfect for breaking down the tar and other poisons from an Exxon Valdez or BP Gulf of Mexico disaster. The problems start if such a device goes wrong. What if the program becomes corrupted and the "nanobots" released to clean up an oil spill start to attack all carbon-based molecules, including living things in the ocean, all the while making more and more copies of themselves? "We cannot afford certain kinds of accidents with replicating assemblers," wrote Drexler in 1986. Nanobots whose instructions mutate, or become corrupted, might end up consuming not just the toxins they were meant to attack but the wider environment in a process that's been called ecophagy. This is the dreaded grey goo scenario: the reduction of everything – the eating of the biosphere and perhaps the entire planet – by ravenous, mutant nano-von Neumann replicators.

Figure 5 Artist's impression of San Francisco in the grip of a grey goo attack. Credit: Jeff Darling.

Bearing in mind such a risk, however remote, it seems likely that nano-devices aimed at cleaning up oil spills and other big tasks won't be allowed to reproduce on their own. Instead the chances are they'll be manufactured in large quantities by "nanofactories" – pieces of equipment that might be small enough to fit on a table-top, but whose fabricating machinery would be inert if removed or unplugged.

Still, there are dangers. Just as some people are both ingenious and irresponsible enough to release computer viruses onto the Internet, so some personality types might be drawn to the challenge of building nano-scale von Neumann self-replicators. More likely a bigger threat would come from rogue nations and terrorist groups who developed the technology as a tool for blackmail. A single grey goo outbreak in, say, New York or Washington, DC, would be disastrous, causing widespread disruption. The entire affected area would have to be isolated and treated just as if there'd been an outbreak of Ebola or the plague. The situation would be even worse if the weapon were deployed in the atmosphere or the ocean, where it would be hard to limit its spread. A single replicator that was able to make a copy of itself every fifteen minutes or so could result in a population of tens of billions of replicators after just ten hours; within two days, assuming it could find enough material to consume, the von Neumann swarm would be as massive as the Earth; in fact, it would *be* the Earth – transformed entirely into goo.

Perhaps this fate has already befallen other planets in space. Their inhabitants, just a bit more advanced than ourselves, may have looked forward to a bright, nano-engineered tomorrow in which molecular manufacturing would lead to more efficient solar panels, miracles of medicine, superfast computers, and so forth. But then it all went horribly wrong. Either by accident or malicious design, self-replicating nanobots were set free and went AWOL, eating their designers and their designers' worlds. If this is one of the ways that technological races sometimes meet their ends, those seeking evidence of extraterrestrial intelligence might think about tuning their telescopes to the spectral signatures, not of rocky worlds like the Earth, but of balls of grey goo.

SURVIVAL TIPS

Don't breathe in or ingest harmful nanoparticles would be the ideal advice, but it is not exactly practical since these specks of manufactured matter are invisible to the human eye and may soon be pretty widespread in the environment. We'll just have to hope that in the absence of strict regulations governing what new nanomaterials can and can't be let loose into the surroundings, companies are cautious about what they expose us to. With luck, the benefits of nanotechnology will far outweigh the risks. But if ever a global grey goo outbreak occurs, heading for the hills probably won't help since they'd be turned to goo as well. If ever the ultimate peril of nanotechnology overwhelms the Earth, the only chance of survival for the human race will be if colonies have been set up on worlds that might survive the attack.

CHAPTER 2

WHEN PHYSICS GOES WRONG

In a circular tunnel twenty-seven kilometers in circumference buried on average a hundred meters below ground level on the French-Swiss border, near Geneva, is the world's biggest experiment. Some people fear it's also the world's most dangerous experiment, capable of destroying not only the planet on which we live but the entire known universe.

The Large Hadron Collider (LHC) is the most powerful particle accelerator ever built, capable of ramming together two beams of protons (moving in opposite directions) so violently that each proton has an energy of up to seven trillion electron-volts (TeV). In everyday terms, that means, when the LHC is cranked up to maximum power, each of its accelerated protons will have about as much kinetic energy (energy due to movement) as a bumblebee flying at top speed. Considering that a proton is about a trillion times smaller across than a bee, that's a pretty impressive amount. Almost unbelievably, the total kinetic energy of a beam of protons in the LHC when at full power will be roughly equivalent to the energy of a *Nimitz*-class aircraft carrier (an 88,000-ton naval behemoth) moving at about ten kilometers per hour. At their fastest, protons in the LHC will hurtle along at 99.9999991 percent of the speed of light, which means they could reach the Moon in just over a second. Because of some early gremlins, the LHC won't be dialed up to maximum until 2014, giving us plenty of time to speculate what might happen when it's running at full throttle.

So far the LHC has survived several lawsuits aimed at shutting it down, and Earth along with the rest of the space-time continuum has survived the LHC. But as physicists gradually ramp up the power of this mightiest of particle colliders, the doom-mongers

Figure 6 The ATLAS detector at the Large Hadron Collider. Notice the size of the person standing at lower centre for scale. Credit: CERN © 2011.

are refusing to go away and fiction writers continue to spin far-fetched tales based on this most awesome of scientific devices. The key question is whether the experts can be trusted, or is the LHC really on the brink of triggering Armageddon?

IN SEARCH OF THE GOD PARTICLE

Like the Hubble Space Telescope and the International Space Station, the Large Hadron Collider is an attention-grabbing piece of extreme engineering with a price tag to match – in the region of six billion dollars. Like Hubble but unlike the International Space Station, it also promises a scientific revolution. One of the *raisons d'être* of the LHC is to test the existence of a curious object called the Higgs boson, sometimes referred to as the "God particle" (an unfortunate nickname coined by the American physicist Leon Lederman),[10] which may or may not underpin our future understanding of how the universe is the way it is. Another is to hunt for new particles predicted by a theory known as supersymmetry (or SUSY, pronounced "suzie", to those in the trade).

Physicists have painstakingly devised what they call the Standard Model to bring order to the world at the subatomic level. In this cunning scheme there are currently twelve basic particles of matter and another four particles that act as force carriers. Of the familiar particles that make up atoms – protons and neutrons in the nucleus with electrons whizzing around in orbits outside – only electrons are truly indivisible or elementary. Protons and neutrons, experiments have shown, are made up of even smaller particles called quarks, of which in the Standard Model there are six distinct varieties, whimsically tagged "up", "down", "top", "bottom", "charm", and "strange". A proton, for example, is really a bag of three quarks, two up and one down; a neutron is made of two downs and one up.

The six quarks comprise a subfamily within the Standard Model. The electron is a member of another six-particle grouping called the leptons, which means "lightweight" (from the Greek *leptos*). A big difference between these two subfamilies of elementary particles of matter – the quarks and the leptons – has to do with the way they interact through the basic forces of nature, of which there are four. Two of the fundamental forces, gravity and electromagnetism, are familiar from the everyday world because they operate over large distances. The other two, known as the strong force and the weak force, are important only over tiny distances, such as those within an atom, so we're not normally aware of them. Both quarks and leptons can "feel" the weak force (as well as gravity and electromagnetism), but only quarks feel – in other words, can interact with one another through – the strong force.

Both quarks and leptons are classified as fermions. A defining feature of fermions is that their spin – an important internal property of subatomic particles (but not really spin in the normal sense of whizzing around on an axis) – can take only half-whole-number values. The spin of all quarks and leptons is ½ .

The four fundamental forces – gravity, electromagnetism, and the strong and weak forces – are conveyed by their own subfamily of particles known as exchange particles. All the exchange particles are bosons, which form a fundamentally different group from the fermions. The spin of bosons is always a whole number.

Best known of the exchange particles is the photon, which carries the electromagnetic force. When two charged objects interact, the electromagnetic force between them can be thought of, at the subatomic level, as arising from the rapid exchange, to and fro, of photons. In the case of charged particles which have like charges, so that they push each other apart, we can picture this as being like two ice skaters who toss a heavy ball back and forth – each throw and catch causing the distance between the skaters to increase. But the comparison isn't really physically accurate, and there's no comparable analogy to help understand how exchange particles give rise to attraction: we simply have to accept that this is how nature works in the wildly unfamiliar realm of the extremely small. Just as electromagnetism has the photon, so the strong and weak forces have their own exchange particles – gluons for the strong force and the W- and Z-bosons for the weak force. Gravity is also assumed to have an exchange particle, known as the graviton, though this has yet to be observed because the equipment would have to be vastly more sensitive.

The Standard Model also includes another particle, which hasn't so far been detected. This is the Higgs boson – the "God particle" – whose possible existence was first mooted by the English physicist Peter Higgs in 1964. One of the outstanding puzzles of physics is why fundamental particles have the masses that they do, or, to be frank, why they have any mass at all. Postulating the Higgs boson is one way of explaining the masses of the quarks and leptons, and also why the W- and Z-bosons have mass, whereas the photon and gluon don't. Other theories have been suggested in which the Higgs isn't called for. The only way to settle the issue is to look for the Higgs, but that demands smashing particles together at super-high energies, higher than early generations of particle accelerator could manage.

A more fundamental question is why there are different types of particles (fermions and bosons) and forces. In a true "theory of everything" the subatomic zoo would reduce, at high enough energies, to a single animal. This holy grail of physicists, which would also see the unification of quantum mechanics and the general theory of relativity, probably lies a long way off. But a first step towards it might be a grander scheme of nature, beyond the Standard Model, known as supersymmetry. One of the predictions of supersymmetry is that for every fermion and boson there ought to be a "superpartner" – a squark for the quark, a selectron for the electron, and so on. Another goal of the Large Hadron Collider, then, is to look for signs of these superpartners amid the shrapnel from high-energy collisions.

To have a good chance of spotting the Higgs, or any of the suspected superpartners, physicists have to arrange for conditions that haven't existed in the Universe since a split second after the Big Bang *and in a form that can be closely studied*. At earlier and earlier times, the Universe was an increasingly simpler place in terms of the variety of stuff it contained. Particle physicists deal a lot with the concept of symmetry. When the cosmos was brand new it was highly symmetric in the sense that there was no difference in strength between the four fundamental forces. Then, an inconceivably tiny moment after time itself began, gravity broke away from the other three forces, like a crystal condensing out of a solution as the solution cooled. Later, as the cosmic temperature fell further (though it was still fantastically high), the strong force declared independence, and finally the symmetry between the weak force and electromagnetism was broken.

If supersymmetry is a fact of nature then it too must have been broken at some point well within the first microsecond of cosmic time. Before that point, fermions and bosons would have existed alongside their superpartners in equal numbers, but after the breaking of supersymmetry the superpartners would have disappeared. The only way to conjure them back into reality – if they ever existed – is to recreate for human observation the ferociously hot, esoteric soup of particles, known as a quark-gluon plasma, which filled the Universe in that long-ago era. Hence the need for the LHC: a genesis machine designed to restage, in a sense, the first few moments of the cosmos.

FAST FACTS

Think microscope, fridge, and vacuum chamber all rolled into one; then think the biggest and most extreme of all these things and a picture of the LHC begins to emerge. It may seem odd that it takes the largest instrument ever made to find out more about the smallest objects in nature. But high energies are the key to unlocking the secrets of nature at its finest level, and repeatedly nudging charged particles to higher and higher speeds with powerful magnets is the only way to reach those energies. The faster a particle travels, the harder it is to bend its path into a circular shape. To reach speeds of 99.999999 percent of light-speed, even using some of the strongest magnets ever built, demands an enormous turning circle, which explains the LHC's extravagant size. The magnets themselves are as big as the trunks of mighty oaks, and hollow so that the particles being accelerated can pass right through the middle of them.

Underlying the strength of the LHC's magnets is a bizarre effect known as superconductivity, which can only be understood in terms of quantum mechanics. When some substances are cooled way down, they become able to pass an electric current – a flow of electrons – without any resistance at all. This means it's possible to make a hugely powerful electromagnet using coils of superconducting wire, because a very large current can be passed through the wire (without any losses through heat), which in turn gives rise to an intense magnetic field.

The niobium-titanium wires at the heart of the LHC magnets are chilled to within a couple of degrees of absolute zero – the lowest temperature that's theoretically possible. To maintain this level of super-frigidity throughout the approximately 7,000 magnets spaced around the LHC's vast doughnut-shaped track calls for about 100 metric tons of liquid helium.

Failure of a superconducting magnet, when part of its coil system jumps back to a normal (high) state of resistance, is known as a "quench". This can happen when the magnetic field strength gets too high (which destroys the superconductive state), or variations in the field strength are too large (which causes local hot spots, which in turn cause the breakdown of superconductivity). Just over a week into start-up operations of the LHC, in 2008, a section of about a hundred magnets unexpectedly quenched leading to an immediate shut-down of the whole collider. About six metric tons of liquid helium vented into the main collider tunnel, the local fire brigade was called out as a precaution, an enquiry followed, and there was a delay of more than a year before the LHC returned to the business for which it was intended – smashing together subatomic particles at record-high energies.

Faster and faster the particles – protons – race around the collider, in two oppositely spinning beams, until they're traveling so fast that they can complete over 11,000 circuits of the twenty-seven-kilometer-wide ring each second. Nothing must get in their way until the time is right for the two beams to collide, and for this reason the internal pressure is kept at one ten-trillionth of an atmosphere – ten times less than the near perfect emptiness on the Moon and as good a vacuum as nature has to offer anywhere in the Solar System.

When the proton beams are eventually allowed to crash into each another, at any one of four principal experimental stations located around the main ring, the result is up to 600 million collisions per second and spot temperatures 100,000 times higher than those in the Sun's core. Needless to say, the LHC is a power-hungry beast, consuming enough electricity to run 120,000 California-style homes.

The four main detectors (plus two smaller ones) where collisions take place, and the outcome registered, are housed in vast underground chambers, as big as cathedrals. Two of these detectors, called ATLAS and CMS, are general-purpose instruments that will spearhead the hunt for the Higgs and supersymmetry particles. The other two, known as ALICE and LHCb, will be used to carry out detailed studies of the quark-gluon plasma mentioned earlier, and also to seek clues to what happened to all the antimatter that was supposedly created along with ordinary matter when the Universe was new. The four major experiments, when in full cry, will spawn a staggering fifteen million gigabytes of data every year, or roughly 1,000 times as much information as is contained in all the new books printed annually. To crunch all that data calls for computing power on an epic scale, and is provided by thousands of computers and storage systems scattered across hundreds of data centers in thirty-four countries, making up what's called the Worldwide LHC Computing Grid.

INTO THE UNKNOWN

The wonderful thing about the Large Hadron Collider is that whatever it discovers it will revolutionize science. Finding the Higgs, for instance, will be stupendous because it will complete the Standard Model and solve one of the great outstanding problems in physics – why elementary particles have mass. Not finding the Higgs will tell us that the Universe is a very different animal to what most scientists suspect. The same applies to supersymmetry particles, and to other potential disclosures. Positive and negative outcomes alike will be fascinating and deeply revealing.

But the fact that the LHC will open up new realms of science, and probe unprecedented regimes of high energy, also means it has the potential to take scientists completely by surprise. Some of the entities it gives rise to in its Big Bang-mimicking smash-ups may be exotic in the extreme – and disturbingly unpredictable.

A very real possibility is that the LHC will manufacture black holes. The mere mention of these may conjure up images of star-eating behemoths that stalk the Galaxy gobbling up everything in their path – not the sort of object you'd want to have roaming around anywhere near your family or terrestrial backyard. But the kind of black holes that the LHC could conceivably spawn are puny in the extreme, weighing less than a dust mote and measuring a mind-bogglingly tiny ten trillion trillion trillionths of a meter across. Conventional wisdom says that black holes this small will effectively disappear almost the instant they're formed – evaporating in a puff of gamma rays through a quantum process known as Hawking radiation. They wouldn't have time to swallow a solitary proton let alone begin dining on our planet's innards. On the other hand, no one has ever seen Hawking radiation and it may be that micro black holes are more stable than most physicists have been led to believe.

Much depends on the particular mathematical scheme of space and time used to run the calculations. Physicists Roberto Casadio, of the University of Bologna, and Sergio Fabi and Benjamin Harms, of the University of Alabama, published some interesting results in 2009 based on a model in which the normal four dimensions of space and time are embedded in a fifth dimension of space (the so-called Randall-Sundrum brane-world scenario).[11] Their work suggested that micro black holes might hang around much longer than the paltry million trillion trillion trillionth of a second normally quoted. It would then be a contest between how fast the holes evaporated and how fast they could gulp down matter from their surroundings. Despite raising the specter of long-lived micro holes, Casadio and his chums were upbeat about our survival chances, concluding that "the growth of black holes to catastrophic size does not seem possible. Nonetheless, it remains true that the expected decay times are much longer (and possibly much greater than 1 second) than is typically predicted by other

Figure 7 Artist's impression of an artificial black hole, created by accident, in the process of devouring the Earth. Credit: Jeff Darling.

models." Despite their comforting words, Fox News was quick to jump on the story and make it seem just that bit scarier. "If the worst comes to pass," said the Fox piece, "and there's now a slightly greater chance that it might, at least it might explain why we've never heard from extraterrestrial civilizations: maybe they built Large Hadron Colliders of their own."

STRANGERS AND THE NIGHT

Earth-eating micro black holes aren't the only hypothesized offspring of the demon LHC seemingly designed to keep nervous folk awake at night. In another scary imagining, the European collider would give birth to strangelets – bound states of roughly equal numbers of up, down, and strange quarks. "Strange," it must be remembered, in particle physics speak doesn't equate to "weird" or "bizarre" in ordinary English; it's just a memorable but meaningless label (like "charm", "colour", or "beauty") for a property that some subatomic particles happen to have. Strange particles certainly exist and are well known in high-energy experiments, but strangelets, which would be bigger and would have the mass of small atomic nuclei, remain purely hypothetical. *If* they prove to be real and *if* they were ever manufactured on Earth, however, they could spell trouble.

The most striking thing about strange matter is that in the fleeting forms of it with which physicists are familiar it's very unstable. Every strange particle that's ever been created in an experiment on Earth to date has contained no more than one strange quark and has decayed in less than a billionth of a second. In theory something different might happen if a particle, or a piece of matter, had a bigger proportion of strange quarks: when the strange quarks started to rival the non-strange quarks in number, the strange particle comprised of them ought to become much more stable. Strangelets are the smallest bits of strange matter that could survive for a relatively long time. But theory suggests there could be much bigger objects made of strange matter, up to and including entire "strange stars", more commonly known as quark stars. These, it's been suggested, might be the final evolutionary state of some stars whose mass is close to, but slightly less than, the critical amount needed to cause them to collapse to become black holes.

Theory also suggests that if strangelets ever came into contact with ordinary matter they'd begin converting it into strange matter in a way analogous to how a seed crystal can trigger a mass outbreak of crystallization in a saturated solution. The original strangelet would bump into a nucleus causing it to change spontaneously into a strange nucleus; the enlarged and more stable strangelet would then assimilate another nucleus, and so the process would go on, spreading and accelerating until, in an inconveniently short time, the whole Earth would be transformed into a big, hot ball of strange matter. Needless to say, we wouldn't survive the metamorphosis.

The danger from strangelets was first raised in connection with another high-energy device called the Relativistic Heavy Ion Collider (RHIC) at Brookhaven National Laboratory on Long Island, New York.[12] The RHIC smashes together gold nuclei moving close to the speed of light, and there were concerns before it went into action over a decade ago that it might prove to be a strangelet factory with unpleasant consequences for the planet. A couple of academic papers appeared at the time which took the issue seriously, but decided, after careful analysis, that there really wasn't anything to worry about. In fact, the RHIC has been operating since 2000 without any noticeable nasty side effects. It turns out that accelerators like the RHIC which run at lower energies than the LHC are better candidates for making strangelets, so that, on this score at least, the European monster machine seems to represent no threat at all.

Slightly more worrying is the possibility of the LHC bringing about a dreaded "vacuum metastability event". One of the big ideas that emerges from quantum field theory – the underlying paperwork of particle physics – is that the Universe and everything in it isn't in the most settled state it could be. In other words, what might seem to be a region of completely empty space in our universe, or a perfect vacuum devoid of even a single particle or speck of energy, is in reality a "false vacuum".

A false vacuum gives the illusion of emptiness and isn't necessarily a permanent state of affairs. Physicists use the word "metastable" to describe something that looks stable but has the ability to flip to a more stable state (see figure 8). So "vacuum metastability event" means a disturbance that causes a false vacuum – a metastable system – to flip into a lower, true vacuum state.

It's as if we lived in a valley hemmed in by tall mountains and thought that our valley was the lowest there was, but in fact, over the mountains, hidden from view, is a lower one. Where the analogy breaks down is that being in a higher valley on Earth isn't dangerous, but being in a Universe that's in a higher-than-zero energy state is a bit precarious. If ever the barrier separating the false vacuum from the true vacuum were to be breached, then our entire cosmos and all of its contents – including the Sun, Earth, and ourselves – would wink out of existence to be replaced by an alien reality with its own set of physical laws and conditions that would almost certainly preclude any kind of life.

One way this end-of-the-Universe scenario could kick in is naturally, due to some event somewhere, somehow, across the vastness of space, that creates a shortcut or passageway between the local minimum of the false vacuum and the absolute, lowest possible minimum of the true vacuum. This effect, known as quantum tunneling, happens when a particle, instead of going over an otherwise insurmountable energy barrier, tunnels its way through. Quantum tunneling is seen in radioactive decay (notably that involving the emission of alpha particles) and is exploited as an essential mechanism in devices such as the scanning tunneling microscope. In the case of the false vacuum, the effect would spread out from its source at the speed of light and eventually reach our planetary system and erase us in the blink of an eye from the scheme of things. Another way it could happen is by an artificial zap that focused enough energy in one tiny place so as to penetrate the barrier separating us from the true vacuum. The LHC, it's been suggested, or some even more powerful successor,

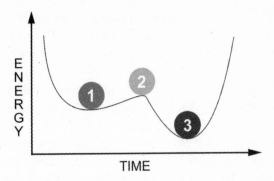

Figure 8 If the Universe is in a metastable state (1) then it might be possible for it to be tipped over the lip (2) of its local energy "crater" into a deeper energy minimum (a true vacuum state) (3) – with disastrous consequences for us. Credit: Jeff Darling, adapted from a Wikipedia diagram.

might supply the spark to tip life, the Universe, and everything into the ultimate oblivion.

BACK TO THE FUTURE

What gives the public at large scope for worrying about these things is that scientists themselves – a handful of them, at least – have raised concerns about the LHC and other high-power colliders, thereby putting a kind of official stamp of approval on the various doomsday scenarios. A couple of these "expert" analyses have also taken seriously the idea that the LHC could work as a kind of time machine.

Irina Aref'eva and Igor Volovich, mathematical physicists at the Steklov Mathematical Institute in Moscow, speculated early in 2008, before the LHC fired its opening salvos, that it might give rise to wormholes.[13] These are like black holes but with associated passageways, or tunnels, which could lead to other points in space and time. The Russians argued that the LHC would focus enough energy in specific places that it might set up "closed timelike curves" – the kind that, in theory, could make time travel possible. A wormhole, if it existed, would play havoc with the normal order of things. One end might be here and now, while the other was anywhere else in space or time, including the past or future. Effectively the wormhole would serve as a shortcut through space-time, something like the fictional "subspace" of *Star Trek*. Aref'eva and Volovich suggested that the LHC could unwittingly become the first ever time machine, providing future time travelers with a stepping off point for journeys into their past.

At first sight, having visitors from the future doesn't sound so catastrophic. If we could learn the trick we might travel back to any point of our choosing. Time travel agents would be able to offer trips to the Cretaceous to "walk with the dinosaurs", or city breaks in ancient Rome and Athens, with a chance to meet Julius Caesar perhaps or Alexander the Great. But, joking aside, trekking into the land of yesterday could prove to be a problem depending on what happens when you get back. The old chestnut of the time traveler who hops into the past and accidentally kills his grandfather illuminates an interesting point: if it's possible to change the past, what happens to us – and the rest of the world – when we return to the present? It may be that, in such circumstances, even tiny changes to the past are amplified and end up rewriting history. This was the theme of Ray Bradbury's 1952 short story "A Sound of Thunder",[14] in which the killing of a butterfly during the time of the dinosaurs causes the future to change in subtle but disturbingly noticeable ways. Alternatively, it might be that by tampering with the past we'd end up creating a fork in time – down one branch would be the future as it was before our trip, down another the future as it had been rewritten. Either way, the consequences could be dramatic: we might end up changing or even destroying the world as we knew it.

The idea of jumps through space-time using wormholes goes back to 1988 when physicist Kip Thorne and a couple of his graduate students at the California Institute of Technology, Pasadena, took up Carl Sagan's challenge of devising a credible scientific basis for more-or-less instantaneous interstellar travel.[15] Their theory, in a form palatable to the lay reader, found its way into Sagan's novel *Contact*,[16] which was eventually turned into a movie starring Jodie Foster.

Unfortunately – or fortunately, depending on how you look at it – the kind of wormholes that Aref'eva and Volovich envisaged the LHC producing would be of the ultra-Lilliputian variety, submicroscopic in dimensions, and not the sort that would be handy for humans to squeeze through. The most we could probably hope to see, the researchers thought, would be a signature that wormholes had popped into existence, perhaps in the form of energy having gone missing because some of the particles manufactured in collisions had disappeared from the here-and-now down a space-time plughole.

LHC time travel derring-do of a different and even more mind-bending type was entertained in 2009 by Holger Nielsen, of the Niels Bohr Institute in Copenhagen, and Masao Ninomiya, of the Yukawa Institute for Theoretical Physics in Kyoto.[17] Their suggestion came in the wake of a series of setbacks that delayed the accelerator coming on-line. The most serious of these incidents happened during tests in September 2008 when the electrical connection between a pair of the collider's giant superconducting magnets vaporized, causing a massive leak of liquid helium and the "quench" described earlier. As a result, all of the thousands of such connections throughout the instrument had to be checked and upgraded, leading to months of delay and a decision to run the LHC at only partial power, with staged increases, for the next few years. Nielsen and Ninomiya's explanation of these "accidents" was that the Universe we live in is so incompatible with the existence of the Higgs boson that any attempt to create this subatomic monstrosity will inevitably cause the attempt to fail. They proposed a cosmic censorship rule which, coming into effect anytime a Higgs threatened to appear, would send ripples back through time to disrupt the event or device that led to the particle's creation.

Not only the LHC fell foul of the cosmic no-Higgs dictum in the Nielsen–Ninomiya scheme of things. In the 1980s and 1990s, American physicists had hoped to build a giant machine called the Superconducting Super Collider (SSC). With an accelerating ring eighty-seven kilometers in diameter and a top energy of 20 TeV per proton beam, the SSC would have dwarfed the LHC. Construction began near Waxahachie, Texas: seventeen shafts were sunk and more than twenty-three kilometers of tunnel bored by late 1993. Then the US Congress pulled the plug on funding for the project – an event which Nielsen and Ninomiya describe as "almost a remarkable piece of bad luck". In fact, perhaps not really bad luck at, the physicists suggest, but, considering that the SSC would have been the first device potentially capable of creating the infamous God particle, its demise was due to the effects of the nature-hates-the-Higgs rule subtly influencing the minds of American politicians.

Needless to say, Nielsen and Ninomiya's preprint wasn't given a heroes' reception by fellow scholars.[18] Perhaps it's just that professional physicists aren't imaginative enough, or perhaps they objected to the authors' somewhat tenuous grasp of English grammar. In any event, the theory didn't emerge well from the peer review process.

But that harsh judgment may have been premature. Fast-forward to 2010. On the UK gadget blog Crave, news broke of a curious incident at the European high-energy lab:

A would-be saboteur arrested today at the Large Hadron Collider in Switzerland made the bizarre claim that he was from the future. Eloi Cole, a strangely dressed young man, said that he had travelled back in time to prevent the LHC from destroying the world.

The LHC successfully collided particles at record force earlier this week, a milestone Mr Cole was attempting to disrupt by stopping supplies of Mountain Dew to the experiment's vending machines. Mr Cole was seized by Swiss police after CERN security guards spotted him rooting around in bins. He explained that he was looking for fuel for his 'time machine power unit', a device that resembled a kitchen blender.

Police said Mr Cole, who was wearing a bow tie and rather too much tweed for his age, would not reveal his country of origin. "Countries do not exist where I am from. The discovery of the Higgs boson led to limitless power, the elimination of poverty and Kit-Kats for everyone. It is a communist chocolate hellhole and I'm here to stop it ever happening."[19]

Some readers – presumably still short of their first morning coffee – were taken in by the yarn. But others quickly spotted the spoof, and the date – April 1.

BEEN THERE, DONE THAT

April Fool jokes aside, some worried citizens have been so deeply moved by what they see as the LHC's capacity for devastation that they've sued CERN – the European organization that runs the collider. In March 2008, former U.S. nuclear safety officer Walter Wagner and Spanish journalist Luis Sancho filed a lawsuit in Hawaii's US District Court. It called on CERN, along with the US Department of Energy, Fermilab, and the National Science

Foundation, which are also involved with the project, to delay the accelerator's switch-on until its safety could be thoroughly reassessed. A few months later the case was thrown out based on the testimony of several senior physicists who argued that the safety fears were unfounded.

Five years earlier, a group of independent scientists had been mandated by CERN to review the various doomsday scenarios mooted before the machine was even built. In a report issued in 2003, they concluded that, as with other high-energy particle experiments, such as the RHIC and Fermilab, particle collisions at the LHC posed no realistic threat to the public or the environment.[20] In the light of the 2008 lawsuit and continuing rumblings about safety issues in the media, CERN commissioned a second review, which came to the same conclusions.[21] At the heart of the rebuttal was the simple observation that whatever physical conditions and events might occur in the LHC, collisions at vastly greater energies happen routinely high up in the Earth's atmosphere. "Nature," wrote the LHC Safety Assessment Group, "has already conducted the equivalent of about a hundred thousand LHC experimental programmes on Earth – and the planet still exists."

Every day, from every direction, cosmic rays arrive at our planet from deep space. Cosmic rays consist of very fast-moving charged particles – mostly protons (hydrogen nuclei) but also including some helium nuclei, and electrons. Some of them come from the Sun, others from the rest of our Galaxy, and others still from extragalactic sources. Their energies span a wide range and can be prodigious, so much so that in slamming into more-or-less stationary nuclei in our atmosphere the collisions can be at least a million times more violent than any head-on meeting of protons in the LHC. If it were possible for the European particle smasher to spawn micro black holes, strangelets, or vacuum metastability events, then all these things should have been produced time and time again through cosmic-ray collisions over the course of our planet's history. Yet here we are still. It's true that the LHC has the advantage, from a physicist's point of view, that it can generate lots of collisions in a given time, and it can do so in a controlled, predictable way. But any calamities threatening the world and its inhabitants by super-fast subatomic impacts ought to have manifested themselves long before humans appeared on the scene.

NOW YOU SEE IT, NOW YOU DON'T

Scientific instruments themselves might not be the main problem. Curiosity killed the cat, so the saying goes, but it could also kill the cosmos according to physicists Lawrence Krauss and James Dent. Doing nothing more than looking at the Universe, they claim, might already have dramatically sped up its demise.

The potential villain here is dark energy. Discovered only recently, and of a nature still to be determined, dark energy accounts, incredibly, for almost three-quarters of all the energy (including energy in the form of matter) in the Universe. (A further twenty percent is dark matter, leaving only five percent in the familiar guise of ordinary matter.) Also implicated in this horrible vision of a hasty cosmic slide into oblivion is our old friend the false vacuum.

Up until the late 1990s, cosmologists thought that the origin of the Universe happened, in the event called the Big Bang about 13.7 billion years ago, when a primordial bubble of relatively high-energy false vacuum spontaneously "decayed", just as a radioactive nucleus decays, into a zero-energy true vacuum. That would be a reassuring state of affairs because a true vacuum doesn't have anything else it can decay into and therefore would be permanently stable. But then along came dark energy. Or, to be exact, along came observations of ancient stellar explosions – supernovae – which showed that, astonishingly, the expansion of the Universe (known about for many years) wasn't gradually slowing down, as everyone had assumed, but was actually speeding up. This could only mean there was some kind of antigravity effect at work – an unseen entity, which quickly became known as dark energy.

The existence of dark energy suggests that the Big Bang didn't produce a completely safe, stable universe at all, but one that was in a false vacuum state. As we saw earlier, one of the fears about the LHC is that it might trigger a cosmos-wide tipping over of this supposed false vacuum that we're in to a lower-level, genuine vacuum, spelling our doom. In their paper, published in 2007, Krauss and Dent pointed out a more subtle and insidious way this could happen.[22]

Commenting on their work, Krauss said: "incredible as it seems, our detection of dark energy may have reduced the life expectancy of the Universe." You could be forgiven for missing that sensational conclusion from a casual glance at Krauss and Dent's highly technical paper, titled "The Late Time Behavior of False Vacuum Decay: Possible Implications for Cosmology and Metastable Inflating States". The work is based on an obscure line of research that stemmed from some theory worked out by the Soviet physicist Leonid Khalfin in the 1950s. What Khalfin found is that, over very long periods of time, the behavior of a system that obeys the rules of quantum mechanics *and* is in a metastable state changes. Think about a collection of radioactive nuclei, such as those of uranium. As time passes, these radioactive nuclei will break apart or decay into smaller, more stable nuclei, and they do this according to what's called an exponential law. Without going into mathematical detail, the point is that exponential laws are extreme. They tend to describe situations, like Ponzi schemes or bacterial populations, that are capable of dramatic rates of growth or of change in general. Khalfin showed that in quantum systems, given enough time, the behavior stops being exponential and flips over, abruptly, to a radically different kind of behavior governed by what's known as a power law. Again, you don't need a math degree to grasp the key point which is that exponential beats power; in other words, for the period over which it operates, an exponential law far outstrips a power law in terms of the chances or the rate of something happening.

The reason why all metastable quantum systems, like collections of radioactive nuclei, studied in the lab are seen to behave exponentially is that they decay long before there's even a remote probability of them flipping over into power-law mode. But the situation is different, Krauss and Dent pointed out, when it comes to the Universe as a whole, which is extremely long-lived.

If the current false vacuum state that the Universe is in survives past a certain point – the point at which the decay flips from an exponential to a power-law type – then it should last forever. The reason for this is straightforward. If the false vacuum state continues to grow in size exponentially for all time, but its chances of decay suddenly start following a power law, then the false vacuum will grow faster than it can possibly be destroyed by decay and it will never come to an end. According to Krauss and Dent's calculations, the closer the false vacuum energy of the cosmos is to zero, the less time is needed before the decay rate switches from fast (exponential) to slow (power law). Given the fact that, in our universe, the vacuum energy appears to be just above zero, we could already be well past the point at which the fateful transition took place.

So far, so good. But now we inquisitive humans have started making observations and taking measurements of things far away and long ago, and in the looking-glass world of the ultra-small, when you measure something – anything – you influence how events turn out. The crux of the matter is something called the quantum Zeno effect, which is the subatomic version of the old adage "a watched pot never boils." Suppose there's a quantum system that starts off in a higher-energy ("excited") state X. Unwatched and left to its own devices, it will eventually decay to a low-energy state Y; in the meantime it exists effectively in two ghostly and unseen possibilities at once (a "superposition" of states) – X and Y. If we take a peek at the system early on, there's a high probability that we'll observe it to be still in state X: by making a measurement, we'll have effectively reset the system's internal quantum clock. Krauss and Dent suggest that this is exactly what we may have done, on a cosmic scale, by making observations of supernovae, back in 1998, which led to the discovery of dark energy. We may have reset the false vacuum's decay clock to zero, switching it back to the fast decay regime, and greatly shortening the odds that the Universe will survive.

Sometimes it just doesn't pay to be too curious.

SURVIVAL TIPS

Large, powerful scientific instruments, like the Large Hadron Collider, are designed to explore new regimes of nature that wouldn't otherwise be available for us to study. They're meant to probe the unknown and come up with data that either support or refute existing paradigms in physics. So, inevitably, they can produce surprises – they'd be a waste of money if they couldn't – and there'll always be a small element of risk in operating them. The most that scientists can do is thoroughly assess the dangers in advance and build experiments that, to the best of their knowledge, pose no risk to people. Fortunately, instruments such as particle accelerators are among the most closely monitored pieces of hardware on Earth, so that if anything did start to go wrong technicians and engineers, on site twenty-four hours a day, would be aware of it almost immediately and would take swift action to shut the machine down.

CHAPTER 3

THE ENEMY WITHIN

The young man lying in the gutter, his head propped against a car tire in a side alley of a busy street, didn't attract much attention. Even those who noticed him walked on by, assuming him to be drunk or high on drugs. His stomach turned in convulsions and he coughed violently. Bloody froth seeped from his mouth onto his shirt. For a while he lay still before a fresh bout of convulsions racked his body. Finally someone recognized his plight and alerted paramedics who took him to a nearby ER. The doctor on hand couldn't make a clear diagnosis and he was put on a course of antibiotics. Thirty-six hours later he was lying in the morgue.

When the results of the autopsy came back they were stunning, devastating. The cause of death itself was enough to make world headlines – infection by *Yersinia pestis*. The same organism behind the Black Death, which had wiped out half the population of Europe in the fourteenth century, was now at large in New York. To have the plague on the loose in so crowded and populous a city was bad enough. But far worse was the fact that in its first victim this strain had proved resistant to antibiotics. Thousands more people could already have been infected, and there was now nothing left with which to treat them.

BE AFRAID, BE VERY AFRAID

It's the scenario that everyone involved in public healthcare dreads most: a worldwide outbreak of a deadly, highly contagious disease for which there's no effective countermeasure. In recent years we've been edging closer to that unthinkable prospect. The human population has risen past the seven billion mark, poverty levels have gone up in many parts of the world, and international travel is more common than ever, making it easier to spread any potential pathogen quickly from one place on Earth to another, far away. Ominously, the germs responsible for a number of lethal conditions appear to have been getting tougher, more able to repel our best efforts to wipe them out.

Against diseases caused by bacteria, the most effective drugs are antibiotics. These have been one of medicine's miracle cures since the first of them – penicillin – was discovered by Alexander Fleming in 1928. Antibiotics work by attacking bacteria in a variety of ways, such as stopping them from building cell walls or interfering with their internal biochemistry, and the beauty of them is that (except in cases where they spark an allergic reaction) they're reasonably safe for us to have inside our bodies. Some antibiotics act selectively against certain species of bacteria, whereas others, like penicillin, are said to be broad-spectrum.

The trouble is that, over time, bacteria are getting more and more resistant to antibiotics. That's a straightforward outcome of the most basic tenet of evolution: survival of the fittest. Having been around for a few billion years, bacteria have become pretty adept at it. A single type of bacterium comes in a slew of different strains, each having subtle natural genetic variations which distinguish it from the rest of its kin. In addition, bacterial genes are constantly mutating. Some strains will have a genetic make-up that gives them a bit of an edge when it comes to fighting off antibiotic attack. So when susceptible strains encounter antibiotics, they die, while these naturally resistant strains are harder to kill. This means that the next time we're exposed to a particular type of bacterium, it's more likely to be one that's survived a previous antibiotic encounter – in other words, a resistant one. Eventually the strain becomes resistant to a variety of antibiotics, even though they work in slightly different ways.

A strain of bacterium that's evolved a number of resistance genes is popularly known as a superbug. It can stand up to virtually the whole range of drugs normally used to combat a particular disease. Every year, nearly half a million new cases of multidrug-resistant tuberculosis (MDR-TB) are thought to occur worldwide, with India and China having the highest rates. The World Health Organization has reported that around fifty million people worldwide are infected – four-fifths of these cases being resistant to three or more antibiotics.

Antibiotic-resistant organisms have become an important cause of infections picked up in hospital. The reason hospitals are such hotbeds for resistant bugs is that so many different strains are thrown together in one place along with so many doses of different antibiotics that the natural selection process is vastly accelerated. Also, people who've been hospitalized tend to have weakened immune systems and other conditions, such as open wounds, which leave them more vulnerable to infection.

Of special concern is so-called methicillin-resistant *Staphylococcus aureus*, or MRSA. Garden variety Staphylococci, or "staph", are common bacteria that can live happily on the skin or mucous membranes without causing any health issues – in fact, at any one time, more than a quarter of us have them in residence in our noses. The problems start when staph get inside the body, for instance through a cut, because then they can trigger an infection, which may range from very mild to life-threatening. The normal treatment is a course of antibiotics. But *Staphylococcus aureus*, it turns out, is fiendishly adaptable to antibiotic pressure. It was one of the earliest bacteria in which penicillin resistance was found – in 1947, just four years after the drug started to be mass-produced. At the time, methicillin was the antibiotic of choice, but it's since been replaced by oxacillin because of the damage methicillin can cause to the kidneys.

MRSA was first detected in 1961 and is now a regular and unwelcome visitor to hospitals around the world. It's become resistant not just to methicillin but to amoxicillin, penicillin, oxacillin, and many other antibiotics. It's also constantly adapting, so that researchers developing new antibiotics have a hard time keeping ahead of the game.

MRSA is spread by contact. So you could get it by touching another person who has it on the skin, or by touching objects that have the bacteria on them, which is why antibacterial gel dispensers are now found on every hospital corridor and ward, and in many other public places. MRSA infections can spring up around surgical wounds or invasive devices, like catheters or implanted feeding tubes. Rates of infection in hospitals, especially intensive care units, are climbing throughout the world. MRSA is also showing up in healthy people who haven't been hospitalized. Rates of so-called community-associated MRSA, or CA-MRSA, infection are climbing fast, especially among children and young adults. Epidemics have already broken out in the US, the UK, and elsewhere, leading, in the worst cases, to rapidly progressive, potentially fatal illnesses including severe sepsis and necrotizing fasciitis.

Known commonly as flesh-eating disease, necrotizing fasciitis begins with signs of inflammation, like redness, or swollen or hot skin, if the focus of infection lies in the upper skin layers. The skin color then changes to purple or purple-black, and blisters often form that are associated with dying tissue. If the flesh-eating disease involves nerves within the skin, the patient doesn't feel any pain in the affected area. However, those infected typically have a fever and feel very sick. Clear or hemorrhagic bullae (large blisters) may develop, with the fluid eventually turning into a grey, foul-smelling fluid called "dishwater pus".[23] When the bullae burst, dry, black pieces of dead tissues develop at the site of the rupture and are later cast off. Over time, as the bacterial invasion progresses, worsening infection and obvious dead tissue can be seen. The dead tissue extends beyond the skin and into the fat and deeper tissue below. If the flesh-eating disease isn't treated, the bacterial infection will progress rapidly and eventually prove fatal. Death rates above seventy percent have been seen in untreated cases in which no antibiotics were applied.

Of serious concern is the fact that some strains of necrotizing *Staphylococcus aureus* are now resistant to vancomycin, an antibiotic that doctors had previously turned to as a drug of "last resort" – a final line of defense. A handful of other antibiotics have come on stream that can do the job once trusted to vancomycin, but once these have been breached, there'll be nothing left in the medical armory to throw against the infection.

With antibiotic-resistant infections reaching unprecedented levels, the World Health Organization has warned that the situation has reached a critical point. Each year, in the European Union alone, over 25,000 people die of bacterial infections that are impervious to even the newest antibiotics. Without a concerted effort, people could be dealing with the unthinkable prospect of a worldwide spread of untreatable infections.

Another emergent threat is what's called the New Delhi or NDM-1 superbug, which was first found in the UK in 2010 in patients who'd brought it back from countries like India and Pakistan, following visits for medical treatment and cosmetic surgery. This superbug is causing alarm because of its resistance to carbapenem antibiotics – a class of broad-spectrum antibiotics that are among the most powerful available for treating infections that evade other drugs. A team of researchers from Cardiff University took samples of seepage water and public tap water in New Delhi and found that bacteria with the new genetic resistance to antibiotics had contaminated the Indian capital's drinking water supply, meaning millions of people there could already be carriers.

Worryingly, the gene responsible for the drug resistance had spread to bacteria that cause dysentery and cholera, which can easily be passed from person to person via sewage-contaminated drinking water. "In India, this transmission represents a serious problem … ", warned the scientists, "650 million citizens do not have access to a flush toilet and even more probably do not have access to clean water."[24]

Unstoppable staph infections and the new threat of NDM-1 are bad enough, but there are more terrible and virulent diseases which also have the potential to achieve multidrug resistance. Some of these could plunge us into a medieval nightmare, wiping out whole swathes of the population and bringing society to its knees.

RETURN OF THE PLAGUE

It became known as the Black Death: a foul, untreatable illness that left its victims to die in agony. The name came from the black spots it produced on the skin but the symptoms were numerous and horrific. Painful, swollen lymph glands causing buboes, resembling huge blisters, in the groin, armpits, and neck were characteristic and give the disease its modern name – bubonic plague. High fever, severe headache, seizures, bloody vomiting and urination, aching limbs, coughing, delirium, and pain caused by decomposition and decay of the skin while the victim was still alive, added to the torture. In the infamous outbreak of the fourteenth century, an estimated seventy-five million people died. That figure would be devastating enough today but at the time it represented about seventeen percent of the entire world population of 450 million people. Proportionately, Europe was even more badly affected, losing between a third and two-fifths of its population within a couple of decades. In terms of scale, today's equivalent would be the loss of more than a billion men, women, and children – roughly the population of China. During the winter months, the Black Death seemed mysteriously to go away. Now we know why: the fleas, carried by rats, which were the main carriers of the disease, were dormant. Each spring, when the fleas began biting again, the plague resumed its progress. Sadly, people at the time had no inkling of how the Black Death was being spread. One theory was that cats – never a well-trusted animal in those days – were responsible. Consequently, domestic cats were rounded up and killed, tragically removing the very means by which the rats, and their cargoes of infectious fleas, might have been kept in check.

Medieval society never recovered from the devastation of the plague. So many people died that there were serious labor shortages all over Europe. The workers who were left demanded higher wages because of the extra burden that had fallen on them. When landlords refused to pay up, peasant revolts broke out in England, France, Belgium, and Italy. The power of the Church was questioned, too, prayers having proved worthless against the disease, with the result that a new era of political turmoil and philosophical questioning was ushered in. In the long run, given the oppressions of that period, the outcome can be seen as positive. But if the same thing happened today, the results would almost certainly be a profound regression and an uncertain future for modern man.

The Black Death is believed to have been due primarily to bubonic plague. However, there are three different types of plague and they're closely interrelated. Bubonic plague can progress to lethal septicemic plague (an infection of the blood) or spread to the lungs and become pneumonic plague. The pneumonic form is especially contagious because the bacteria can spread in the mucous-laden droplets shot out during coughing or sneezing.

Other outbreaks of plague have happened since the mass culling of the 1300s, among them the Great Plague of London (1665–1666), the Great Plague of Vienna (1679), and, in the US, the epidemic in Los Angeles (1924–1925) – the last occurrence of the disease in a western urban area. Most recently, in 2010, pneumonic plague surfaced in Tibet, affecting five individuals; a year earlier, an outbreak of the disease in a farming town in Qinghai province, China, killed three and infected nine others, prompting authorities to seal off the community of 10,000 people for more than a week.

The plague bacterium *Yersinia pestis* is a slow-ticking time bomb. Although not very many people fall victim to the plague at present compared with bygone years, it's still carried by rats in many parts of the world and so, gradually, the bacterium is mutating. To date, it's developed nothing like the drug-resistance of MRSA or other common human bacterial infections, but its ability to fight antibiotics will inevitably increase. As long as the rat-to-flea cycle, which accounted for historic plague pandemics, continues, there'll be the opportunity for the selection of new traits that could make the organism more virulent.

The island of Madagascar is a case in point. Human incidence of the plague, which became virtually unheard of in Madagascar after the 1930s, resurfaced in 1990 with more than 200 confirmed or suspected cases each year since. In 1995, the first multidrug-resistant strain of *Yersinia pestis* was isolated, and this in a country not well equipped to tackle major threats to public health. The danger is that such a strain might find its way onto an international flight bound for a densely populated city abroad, carried by a person recently infected.

Were a multidrug-resistant form of the plague to end up in a crowded city like London, Beijing, or New York, it could spread like wildfire before anyone knew what was happening. A few days are needed for the symptoms to appear, and unless treatment follows within the next twenty-four hours the disease is almost invariably fatal. If none of the antibiotics normally used to treat the illness worked, there'd be nothing to stop it, except strict quarantine, and that would probably come too late to prevent thousands of infected people from carrying the illness to all parts of the globe.

The terrible loss of life during the Black Death would look small compared to the havoc that a multidrug-resistant form of the plague could wreak nowadays. An infected city could be decimated in a couple of weeks and the death toll worldwide might reach into the billions.

KILLER FLU

Smaller even than bacteria, and on the borderline between life and non-life, are viruses – disease-causing entities that hijack the machinery of living cells in order to make copies of themselves. Because viruses mutate and reproduce so fast, they're masters of adaptation.

Cells have special proteins, called surface receptors, which act like docking ports for hormones and a whole host of other chemicals that give the cell instructions what to do. Viruses have surface proteins by which they, too, can bind to these surface receptors and thereby sneakily gain access to the cell's interior. During a viral attack the body responds by developing antibodies to fight off the virus. At the same time it acquires an elite force of so-called memory cells, equipped specifically to detect and repel the same virus if it ever tries to mount another assault. This explains why we almost never catch the same strain of cold or flu twice – after the first bout we're immune to it.

It also explains why viruses are so good at mutating into new forms. Since the only way they can reproduce is by infecting host cells, they have to be able to evolve faster than the defenses of these cells. The result is a never-ending struggle at the microscopic level. The body retains memory cells that produce antibodies to attack any viruses that have been encountered in the past and that attempt a fresh invasion; in response, viruses develop new surface proteins to which the antibodies can't attach. In the survival stakes, viruses must always stay one step ahead.

Influenza is a good example. Every year there's a flu season in the autumn and winter months. The influenza virus has two main surface proteins, called H and N proteins, which allow it to attack the body's cells. When we're infected (or vaccinated) with flu we develop memory antibodies, called H and N antigens, against these proteins. The antibodies prevent subsequent infection if we're exposed to the same flu virus. But as the flu virus travels around the world following the local flu season, it mutates. The chances are that only one of the surface proteins, H or N, will change. When this happens and we're exposed to the mutated virus during the next year's flu season some of our antibodies will bind to the virus and some won't. Those that bind to the non-mutated protein will still give us some protection. Those antibodies against the mutated protein will no longer work. So, the infection will be mild but will infect a lot of people. This process of gradual mutation from year to year is called antigenic drift and leads to annual flu epidemics. Every year, organizations such as the Health Protection Agency in the UK and the Centers for Disease Control and Prevention in the US keep a close eye on how the flu virus has mutated and make available a new vaccine based on these new strains. So, every year we have the option of having a new flu shot.

The flu virus has another way of evading our immune systems. It can rearrange its genetic material by mixing with a different strain of flu to create a hybrid that has new H and N antigens in the same virus. This can happen when flu viruses from two different species of animal infect the same cell.

Flu viruses tend to be species-specific. Humans, horses, pigs, chickens, and ducks all have their own flu viruses, and there isn't normally any interaction between the different kinds. But if someone, such as a farmer, is in close contact with an animal (for example, through its droppings or nasal mucus) when the animal has flu, the virus can infect the person. This isn't usually a problem, because whatever mild infection ensues doesn't spread to anyone else. But if the infected individual also happens to have a human flu virus at the same time then the two viruses may end up replicating concurrently in the same cells, with dire consequences. The viruses that come out of the affected cells have their genetic material scrambled. Some might be mostly human flu but with animal flu H and N antigens on their surface. This represents a big change – not an antigenic *drift* but an antigenic *shift* – in the virus. It's a rare occurrence, but it does happen, on average every sixteen years or so.

The result of antigenic shifts is completely new flu viruses against which no one has any antibody memory. They're particularly nasty, can spread uncontrollably, infecting a lot of people fast, and give rise to pandemics.

The worst flu pandemic of the last 100 years happened in 1918, and has been called the Great Pandemic or Spanish flu (because one of its first high-profile victims was the King of Spain). Of all the people on Earth (1.6 billion of them at the time), almost a third were infected, and between fifty million and 100 million (up to six percent of the world's population) died from it. In fact, it was one of the major factors in bringing the First World War to a swift end: more soldiers died of flu in the autumn of 1918 than as a result of combat in the entire war. It's been described as "the greatest medical holocaust in history" and may have caused more fatalities than the Black Death.

The Spanish flu struck with terrifying speed, often killing its victims within just hours of the first signs of infection. So fast did the 1918 strain overwhelm the body's natural defenses that the usual cause of death in flu patients – a secondary infection of lethal pneumonia – rarely had a chance to establish itself. Instead, the virus caused an uncontrollable hemorrhaging (escape of blood) that filled the lungs, and patients would drown in their own body fluids. As one medical student at the University of Pennsylvania noted at the time: "After gasping for several hours they became delirious and incontinent, and many died struggling to clear their airways of a blood-tinged froth that sometimes gushed from their nose and mouth. It was a dreadful business."

The origin of the Spanish flu isn't known for sure, but what probably happened is that an avian (bird) flu virus first infected pigs and mixed with a swine flu virus. These viruses exchanged genetic information, leading to the formation of a new virus which then infected humans and began spreading from person to person.

Given that a new strain of flu at least as potent as the Spanish flu could emerge at any time, it's no wonder that government agencies charged with protecting public health monitor closely the way flu viruses mutate from one year to the next. If a *drift* is detected they anticipate a nasty flu season and recommend vaccinations for the very young, elderly, and people with weakened immune systems. If a *shift* is detected then drastic measures are brought in to stave off a pandemic. Sometimes it can seem that these efforts are over the top and that people are being scared for nothing – but bearing in mind the scale of the 1918 disaster, it's better to lean on the side of safety.

In 1976 an antigenic shift was detected from pigs to pig farmers and in the US, President Gerald Ford recommended that everyone in the nation be vaccinated against this swine flu. The pandemic never materialized. Perhaps it was a false alarm or perhaps the massive vaccination campaign nipped the disease in the bud; we don't know. In 1997, several young people in Hong Kong died of flu, and it was determined that there'd been a crossover of genes from chickens. Every chicken in Hong Kong and southern China was slaughtered to prevent the infection becoming established in the human population. Again, no pandemic occurred and the authorities were accused from some quarters of overreacting. But there's no way of knowing whether or not, without the cull, the disease would have mutated into a form that could be transmitted from person to person. It might still do so.

The Hong Kong outbreak was the first case of an avian or bird flu virus known to infect humans directly. Since it happened, the virus responsible, called avian influenza A (H5N1), has spread among domesticated birds across Asia. In 2005 it was found in poultry in Turkey and Romania. The wider the area over which it spreads, the greater the chances of a worldwide outbreak and a major breach of the inter-species barrier into humans. Several hundred cases of human infection with the H5N1 bird flu virus have been reported by more than a dozen countries in Asia, Africa, the Pacific, Europe, and the Near East, most thought to be the result of direct contact with sick or dead infected poultry. The ever-present danger is that the virus will mutate into a form in which it can be transmitted between humans. That would be a disaster because not only would we have no immunity to it or effective treatment for it, but the current death rate for patients with confirmed H5N1 infection is more than fifty percent.

Another concern is that the genes of various potentially catastrophic flu viruses are being harbored in wild aquatic birds (which for the most part aren't affected by them). Periodically, they may make the jump to other species in which they can be far more deadly. The 1918 virus, for instance, is still being maintained in the bird reservoir. So even though these viruses are very old, they have the capacity to evolve, and to acquire new genes and hosts. The potential exists for the Spanish flu to go on the rampage again in the twenty-first century.

NOVEL ENDINGS

Nature also has a habit of conjuring up entirely new lethal diseases. Seemingly out of the blue came HIV (human immunodeficiency virus) which causes AIDS (acquired immunodeficiency syndrome), a condition in which the immune system begins to fail, opening the door to life-threatening infections. Since its discovery in 1981, AIDS has killed more than twenty-five million people worldwide and presently infects about 0.6 percent of the world's population. Sub-Saharan African is particularly badly affected. In countries such as South Africa and Zimbabwe, more than fifteen percent of all adults carry the disease, and the death rate is so high that millions of children have been orphaned.

HIV is now thought to have originated in monkeys in southern Africa and was transferred to humans in the late nineteenth or early twentieth century. It's difficult to treat because it mutates at such incredible speed – so quickly that the immune system never clears it from the body and every vaccine that's been developed has failed because the target is constantly shifting. Although there's no cure for HIV, drugs known as antiretrovirals can delay how long it takes an HIV patient to develop full-blown AIDS. These drugs can significantly extend the life of someone with HIV, but they have to be taken every day and aren't readily available in many places that need them the most.

Research continues around the world to develop an HIV vaccine. Progress is being made, although it's likely to be a number of years before such a treatment is widely available. Meanwhile in those countries worst hit by the disease, the effects have been traumatic. Young adults, in the prime of their working lives, are disproportionately struck down because the virus spreads mainly by sexual transmission. The result is a staggering number of orphans being left to their own devices or cared for by elderly grandparents. Economies and societies in countries with a significant AIDS population have been devastated to the point of collapse.

At present, AIDS doesn't threaten the human race as a whole because it can only be transmitted by blood and sexual contact. The situation would change drastically though if the virus mutated to an airborne form, like the common cold virus. One way this could happen is through gene reassortment – a mutation mechanism that involves the merging and reassortment of genes from different viruses. We've already seen how this happens in the case of flu infections, especially involving pigs, which are ideal mixing vessels for bird, human, and swine viruses. There have also been reports that HIV has mutated with tuberculosis in South Africa, making the disease easier to spread and harder to control. Airborne AIDS remains a distant possibility for the moment but it is an example of how some viruses could become megacatastrophic.

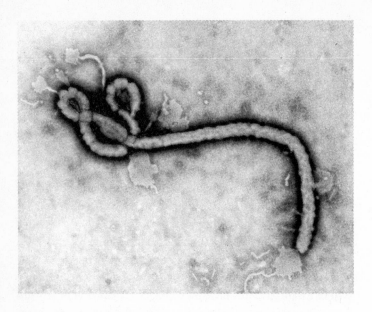

Figure 9 The Ebola virus. Credit: Centers for Disease Control.

Ebola is another viral disease that can make the jump from non-human primates to man, and its symptoms are truly horrific. Flu-like sore throat, headache, and high temperature are followed by nausea, vomiting, and diarrhea, and, in the final stages, massive internal bleeding, either from the major organs or from blood vessels, the digestive tract, and gums. Eventually, so much blood is lost that the patient may go into shock or suffocate because of blocked airways.

Ebola was first identified in 1976 in the Democratic Republic of Congo (previously called Zaire), and in southern Sudan. Compared to AIDS, it's killed only a small number of people to date and very few at all outside Africa. But like several other types of so-called hemorrhagic fever, including Lassa fever and Rift Valley fever, it has the potential to become pandemic, with unthinkable consequences.

The Congolese form of Ebola has one of the highest fatality rates of any human pathogenic virus – about ninety percent of its victims die from it within two to twenty-one days. It spreads to other people who come into contact with the blood and bodily fluids of the infected person, and by contact with contaminated medical equipment such as needles. The high fatality rate, the rapidity of demise (so that people often die before others have chance to catch the virus from them), and the often remote areas where infections occur (making quarantine easier), have so far helped contain the disease. But the possibility of mutation is ever present. If a mutation occurred that allowed a disease like Ebola to spread by airborne infection, international air travel could easily do the rest and bring its horrors to our shores.

Some horrors are already here. Creutzfeldt–Jakob disease (CJD) is a degenerative condition of the brain and nervous system that starts with failing memory, behavioral changes, and lack of coordination, and progresses rapidly until the patient is blind and impaired in many other ways, and finally slips into a coma. CJD is invariably fatal, usually within a year. Seen under a microscope, the brain of a CJD victim is peppered with tiny holes where nerve cells have been destroyed by the build-up of a type of protein called prion protein – a protein whose molecules have become folded the wrong way and therefore dysfunctional.

Prion proteins come in both a normal form, which is harmless and found in the body's cells, and an infectious form. CJD may develop because some of a person's normal prion proteins spontaneously change into the infectious form of the protein, which has a distorted shape and spreads, altering the prion proteins in other cells in a chain reaction. Once they appear, abnormal prion proteins – or "prions" (short for protein infections) as they're called – clump together and start to cause damage to nervous tissue. CJD isn't contagious: it can't be spread from one person to another. Yet there is a way it could strike down many people and have the effect of a pandemic.

Cattle can succumb to a disease which is very similar to CJD called bovine spongiform encephalopathy (BSE), often referred to as "mad cow" disease. A mass outbreak of this happened in the UK in the late 1980s due to the practice of feeding the ground-up carcasses of slaughtered cattle, some of which were infected with BSE, as a food supplement to other cows. The disease then passed to humans, as a new form of CJD called variant CJD (vCJD), when people ate the meat of animals who'd contracted the disease, especially meat containing tissue from the brain and spinal cord.

Tens of thousands of cattle, including entire prize herds, were slaughtered following the outbreak of BSE in Britain. An untold number of people, in the millions, consumed meat that had been contaminated before the cull. Only 170 or so have died of vCJD in the UK to date (and a further fifty elsewhere in the world). But there's great uncertainty about the normal incubation period for this disease: it might be a few years or a few decades, in which case large numbers of people could yet be affected. Studies of a similar disease known as kuru, found in Papua New Guinea, give cause for concern. Kuru is thought to have been transmitted through cannibalism when family members ate the body of a dead relative as a sign of mourning. The practice was made illegal in the 1950s; however, kuru reached epidemic proportions in some Papua New Guinean communities in the late twentieth century, suggesting the incubation period might be thirty to fifty years. If vCJD follows a similar pattern, Britain faces an uncertain future.

HIV, Ebola, CJD, bird flu, and antibiotic-resistant superbugs are just a few of the pathogenic threats to humanity which have surfaced over the past few decades. Our soaring numbers, ubiquitous international travel, and the increasing use of chemicals and biological agents without full knowledge of their consequences, have increased the risk of unstoppable pandemics arising from mutant viruses and their ilk. The plague and Spanish flu are vivid examples from history of how microbial agents can decimate human populations. But the consequences aren't limited to a high body count.

PANDEMICS AND PANDEMONIUM

The Black Death and, more recently, the Spanish flu are stark reminders of the frailty of human society in the face of the tiniest of invaders. It may seem that we're better placed to deal with major pandemics today than in years gone by, thanks to improved medical knowledge and facilities (at least in the developed world). But in some ways we're more vulnerable. The Earth is more crowded and international travel is commonplace so that contagious diseases can spread rapidly to all parts of the planet.

A pandemic would start with mounting rates of illness among all sections of the population. Absenteeism in the workforce, including police, firefighters, paramedics, and healthcare workers, would leave more and more gaps in essential services. Utilities and transport systems would suffer, and supplies of some foods and other products would begin to run out. Quarantines would be imposed, and places of public gatherings such as schools, theatres, and museums would likely be closed in an effort to stem the spread of the disease. National economies would be hit because of restrictions on international travel and the import and export of goods. Healthcare facilities would be overwhelmed. If vaccinations were available they'd become compulsory, and plans might be drawn up for mass burials.

At some point the death toll might climb to the point of disrupting the very fabric of society. According to US government studies, if a global pandemic affecting at least half the world's population were to strike today, health professionals wouldn't be able to cope with the vast numbers of sick and dying. The result of so many deaths would have serious implications for the infrastructure, food supply, and security of twenty-first-century man.

If the pandemic rivaled the scale of the Black Death, health services would collapse. Worse, since many healthcare workers would themselves get sick, it would become impossible to track the spread of the disease. High-maintenance facilities such as nuclear power stations and refineries would go short-staffed and become prone to accidents. Availability of electricity and access to gasoline and other means of transportation would be lost in large areas of the country. With transport badly affected, food shortages and famine would set in, especially in urban areas, possibly triggering riots and revolts over the remaining food supplies. Many people would start to live off the fields as trade and transport facilities ground to a halt. Any remaining government institutions would be paralyzed and slow in responding to emergencies. The result would be widespread lawlessness and mass migrations of people who'd survived the first onslaught of the pandemic.

Mankind would almost certainly survive even a severe pandemic, because if the disease agent were aggressive it would burn itself out in time. Also, quarantine measures would help ensure that some enclaves of the population were protected (unless the incubation time of the disease was really long). However, a pandemic that killed hundreds of millions or more could devastate the world economy and set back human development by many years.

FATAL DNA

While an untreatable pandemic could strike suddenly and potentially bring civilization to its knees in weeks or months, degenerative diseases might do so over longer periods or make it easier for the advance of an infectious disease to be much more devastating. The most common degenerative disease is cancer. Every second man and third woman in the western world will be diagnosed with this disease at some point in their lives (a proportion that, it's true, has been rising in part because we're living longer). Deterioration of our environment through the release of toxins, waste materials, and air pollution, and intake of unhealthy foods is making this problem worse (to different degrees for different types of cancers). If cancer, or some other form of degenerative disease, were to become even more common and strike before reproduction, the very survival of our species could be threatened.

China is now a cancer hotspot because of its rampant and coal-stoked industrialization. It's become one of the most polluted countries on Earth. Statistics from the World Bank reveal the extent of the problem: sixteen of the twenty most polluted cities in the world are Chinese; ninety percent of urban groundwater in China is contaminated; an estimated 900,000 people in this most populous country die every year indirectly from air and water pollution.

Highly toxic heavy metals, such as lead and mercury, contaminate cultivable land, poisoning the food chain, while the number of Chinese who die from lung cancer and other serious respiratory diseases is steadily rising. Instead of clean air many citizens face a choking smog, saturated with carbon monoxide, nitrogen monoxide, sulfur dioxide, arsenic, and lead. In some places, the concentration of fine particles can reach

1,500 micrograms per cubic meter – an extraordinarily high concentration considering that in European countries a figure of forty micrograms per cubic meter of particulates is enough to set alarm bells ringing.

According to the Chinese Ministry of Health, the death rates of lung, liver, intestinal, breast, and bladder cancers have leapt over the past three decades, and all these cancers are closely linked to environmental and lifestyle factors. The numbers for lung cancer (up 465 percent) and breast cancer (up ninety-six percent) are particularly striking.

The same trend, to a lesser degree, is seen in other parts of the world. A study published in the medical journal *The Lancet* estimated that outdoor air pollution is a factor in three percent of lung cancer cases in the UK. Many of these cases occur in workers who spend a lot of time on the road and are exposed to large amounts of diesel exhaust. A recent report in the *Journal of the American Medical Association* also describes increased mortality rates from lung cancer in the most polluted cities, even among lifelong non-smokers. The risk of dying from lung cancer was found to go up eight percent for every additional ten micrograms of fine particles per cubic meter of air.[25]

Cancer-causing compounds occur in petroleum, which contaminates groundwater through seepage or spills. Thousands of underground petrol storage tanks remain in the ground beyond their twenty-year safe life expectancy. Ironically, a petrol additive known as MTBE, used since 1980 to reduce air pollution, has been shown, in laboratory rats and mice that breathe air or drink water contaminated with it, to cause lymphoma, leukemia, and testicular, thyroid, and kidney tumors. On their own carcinogens like these don't pose an existential threat to our species. The real danger is that they might sufficiently weaken our natural defenses, both genetically and in terms of our immune systems, that an additional stress – be it an infectious agent, a parasite, or even a mutation within our own DNA – could prove disastrous for the human population.

Beyond environmentally-related cancer lies the still more frightening prospect of transmissible cancer – a form that could be passed on from one person to another like an infectious disease. Cancers of this type have already been seen in some other animal species.

Most infamous is the Devil facial tumor disease (DFTD), a transmissible parasitic cancer in the Tasmanian devil, a large Australian rodent. DFTD is caused by clones of cancer cells that are transmitted by physical contact, particularly by biting, some feeding practices, and aggressive mating. Once infected by the cancer cells, the Tasmanian devils develop lumps and lesions on the face and body, which interfere with feeding so that the animals effectively starve to death. The population of Tasmanian devils has plummeted by seventy percent since 1996, and as of 2010 about eighty percent of the population was infected with the disease. At least nine strains of the cancer have been identified, showing that it's evolving and may become more virulent. As a result, Tasmanian devils face the very real prospect of extinction in the wild, and there's also the possibility of the disease spreading to other related species, such as the Australian quoll.

Another transmissible cancer is canine transmissible venereal tumor (CTVT), a disease spread by mating in dogs. A single clone of CTVT cells has colonized dogs worldwide, representing the oldest known malignant cell line in continuous propagation. The tumor cells are themselves the infectious agents, and the tumors that form aren't genetically related to the host dog. Although the genome of CTVT is derived from a canid (probably a dog, wolf, or coyote), it now lives essentially as a single-celled, asexually reproducing (but sexually transmitted) pathogen.

If a transmissible cancer were to develop in humans it would be catastrophic – a more deadly killer than any infectious disease or cancer alone. It would combine the speed of transmission of an infectious agent with an inherent weakness of our genome. The reason infectious facial cancer has spread so aggressively through the population of Tasmanian devils is most likely because of a low diversity of immune genes in the DNA code of that species.[26]

What then is the status of our DNA? According to one school of thought, our genetic code is seriously compromised. More than ninety percent of our DNA is considered to be "junk DNA" – chains of code that are derived from viral insertions, duplications, and other inefficiencies. Virus-like components of the human genome make up nearly half of our DNA, while more than a third of our DNA seems to serve no purpose other than to make copies of itself. One study concluded that men may soon be in short supply due to a deterioration of the Y-chromosome.[27] At some point our genome could become so inefficient that it's no longer fit to reproduce accurately. Theoretical studies have shown that it may be very close to the stability/instability boundary with possibly disastrous effects. Beyond this boundary, evolution would begin selecting against us at a cellular level, causing genetic diseases and genetic-related diseases, including cancer, to skyrocket.

A potentially fatal weakness of our DNA, and also that of some of our fellow mammals, is that most of its ability to control cancers depends on a mere 40,000 bytes (to use a computer term) of chromosomal DNA, and in particular a gene known as p53, which is suspected of being directly or indirectly involved in the control of all cancers. A 2005 study involving the exposure of a sample of mice to a highly carcinogenic compound found that seventy percent of the pups born to mice in which the p53 was missing were born with brain tumors, while the offspring of animals with a full complement of the gene were cancer-free.[28]

Of great interest to scientists is a strange-looking creature known as the naked mole rat (or sand puppy), a hairless, burrowing rodent native to East Africa which spends its entire life underground in hierarchical social groups of up to 300 members. Unusually, naked mole rats have not one but two major cancer-protecting genes, and as a result rarely if ever develop this disease. Their normal life expectancy is around thirty years – about seven times as long as that of an ordinary rat. Researchers are keen to learn more about the naked mole rat's resistance to cancer, and how this might help human patients. These efforts will be boosted by the recent announcement that the entire genome of the naked mole rat has now been sequenced.[29]

Don't worry so much about our DNA, argue other scientists. After all, they point out, a lot of species have more non-coding DNA than humans do, including many, if not most, plants and amphibians – and they're able to keep on reproducing without a

Figure 10   The naked mole rat. Credit: Roman Klementschitz/Wikipedia.

problem. Also, the argument goes, men don't need to get anxious about becoming a rarity any time soon because evolution will strongly select against the loss of genes on the Y-chromosome. Perhaps there'll also be a strong natural selection against rampant, uncontrolled cancer. If, for example, a fatal mutation were to infect p53, it ought not to spread far because selection would work against the mutation, eliminating it or giving rise to an alternative protective gene.

So far, at least in humans, cancer remains mostly a post-reproductive disease that doesn't affect the survival of the species. But if somehow cancer were suddenly to become contagious, the threat to us would be amplified and the outcome uncertain.

However the evolutionary wrestling match involving our genes plays out, it seems that our continued presence on this planet hangs by a thread. One subtle change in our genetic make-up – the corruption of a single gene – could leave us at the mercy of our most feared disease. In a carcinogen-ridden world, without this key protection, cancers could begin in the womb, putting the viability of our species in doubt. Or infectious diseases might take advantage of our weakened genetic and immunological defenses and put us on the critically endangered list.

SURVIVAL TIPS

The biggest immediate threat to our survival is from deadly bacterial infections that develop multidrug resistance, or inter-species diseases, such as bird flu, which mutate into a highly contagious form. More effort is needed to stay ahead of the game, in terms of spotting new diseases and strains as early as possible, and developing a range of new antibiotics and other treatments to deal with them.

As far as degradation of our DNA goes, not much can be done about that at the moment. However, as genetic engineering and gene therapy progress, we may learn how to snip out potentially fatal parts of our DNA and insert new genes that may help to protect us against killer diseases such as cancer.

CHAPTER 4

MARCH OF THE MACHINE MIND

No technology is developing more rapidly than that of computers. Every year they become more powerful and capable, and we become more dependent on them. Could computers and robots be on the verge of making us obsolete? Are these inventions of ours about to supersede us or even enslave us?

Many people, still in their middle age, are old enough to remember the dawn of personal computers, back in the mid- to late 1970s. It started with the Altair 8800, and then quickly progressed to such relatively sophisticated models as the Tandy TRS-80, Sinclair ZX, and Commodore 64, with anywhere from 1,000 to a seemingly astounding 64,000 bytes of random-access memory (RAM). By and large, these were no more than games machines, with totally unreliable cassette tape recorders for external storage, the most primitive of graphics, and a not-so-friendly user interface. The fastest computer in the world in 1976 was the Cray-1, capable of a then-blistering 100 million mathematical operations a second – about ten times slower than an iPad – and with a RAM of up to eight megabytes – hundreds of times less than even a cut-price modern laptop. The Internet existed only in embryonic form as a little-known military network called ARPANET. Skype and Facebook, along with modern games devices such as the Xbox Kinect, would have seemed almost miraculous even to the most starry-eyed whiz-kids of that era.

Computer gaming, which once seemed so trivial and nerdy, has flourished into a lucrative global enterprise. In its opening week *Grand Theft Auto IV* took in over $500 million (£300 million) in sales, while in 2010 the total value of the video games market was about $105 billion (£60 billion) – more than the GDP of many countries. Virtual property, too, which exists only as data within the digital landscape of some imaginary online world, is bought and sold for lavish amounts. In 2005 a twenty-three-year-old gamer who glorified under the moniker Deathifier splashed out £13,700 on an island that in reality was just a string of 0's and 1's on a server farm somewhere, a seemingly crazy investment which he comfortably recouped within a year.

But beyond mere economics, some individuals take gaming so seriously that it becomes life-threatening. In 2005 a Chinese online games player killed an opponent (in real life) for stealing his virtual sword, and in the same year a South Korean died of exhaustion after a fifty-hour marathon session of StarCraft in an Internet café. Five years later, also in South Korea, a couple allowed their three-month-old, flesh-and-blood daughter to die of malnutrition at home while they devoted hours in Internet cafés to playing a computer game that involved raising the virtual character of a young girl.

Extreme cases aside, the fact is we're becoming more and more intimately involved with computers, at home, at work, while traveling, for entertainment, social networking, news, banking, shopping, and so on. Interactive games and the Internet are pervading our lives, and the trend is increasing year by year. With so much of our infrastructure, from commerce to communication, now dependent on the world wide information web, attacks on the Internet, or selective portions of it, are becoming increasingly attractive to groups who'd like to disrupt the western economy or, more locally, disenfranchise their citizens from unpleasant home truths.

The threat of cyberterrorism is here to stay, but it almost certainly doesn't have the potential to be megacatastrophic. Computers that control access to nuclear weapons and other sensitive military systems are "air-gapped" – in other words, they're physically, electrically, and electromagnetically separated from any computers or networks that can be got at by outside hackers. Systems that control nuclear reactors, and other key utilities, aren't always as well defended. In 2007, Scott Lunsford, a researcher for IBM's Internet Security Systems, took just a week to penetrate the SCADA (Supervisory Control and Data Acquisition) software used to control a nuclear power station and effectively take over the running of it. Lunsford thought that with further effort, a malicious hacker might have been able to sabotage the power output from the reactor, causing a temporary regional blackout, but not to have triggered a meltdown.

The real threat to us from computers is more insidious. Computer game addiction is already a problem. In South Korea the government is close to adopting a "Cinderella" law banning youngsters from playing online games past midnight. If a bill submitted to parliament is passed it will require South Korean online game companies to cut off services at midnight for users registered as younger than sixteen. The amount of time that people spend playing computer games is rising, in part because the games themselves are becoming more compelling, drawing the player into an environment that seems almost as tangible as the external world. Further down the road, there's a risk of people becoming more and more absorbed into this extraordinary new digital universe, with potentially damaging consequences for their bodies and minds. It seems likely that, for many, as immersive technologies become ever more advanced, the lure of computer-generated worlds will become irresistible. After all, who'd want to lead a humdrum life when it could seem just as real to be a wizard in Middle Earth or a crime-fighting super-hero?

We're also routinely hooked up with computers and each other through the Internet and telecommunications systems. With crude brain-computer links already a reality, it may be only a matter of time before we effectively merge with the machines we've made. Depending on your point of view, and how it all pans out, that could prove a disaster for humanity or a pivotal moment in our evolution.

BRAINS IN SILICON

Back in the 1960s there was much talk of artificial intelligence soon rivaling the human brain, and even earlier a common theme in science fiction was the threat of smart robots taking over the world. In his 1950 series of short stories *I, Robot* (eventually used as the basis of a successful Hollywood film), Isaac Asimov addressed this danger of mechanized mayhem with his Three Laws of Robotics which were designed to ensure that no robot could ever harm a human being. But crazy computers that go out of control and robots on the rampage remain a popular fear, and films like *The Terminator* have kept the prospect of evil brainy machines firmly in circulation.

In the real world, artificial intelligence (AI) has progressed more slowly than many people expected half a century ago. Getting a machine to understand the vagaries of human language, learn from experience, or display genuine creativity has proven a lot trickier than imagined. Nor is the problem of making a machine do some of the things our brains are so adept at helped by the fact that we're still sketchy on many aspects of how the brain works.

But amid the disappointments AI has sprung some interesting surprises. Chess-playing computers began rivaling human chess grandmasters in the 1990s, and in 1997 won a match against a reigning world champion for the first time, when IBM's Deep Blue beat Garry Kasparov 3½–2½. This victory was foreshadowed years before, in 1979, when a computer running a backgammon program beat world champion Luigi Villa 7–1 – the first triumph of machine over man at a game in which strategy, chance, and numerous possible positions were involved. In 1994 a computer called Chinook overcame Marion Tinsley, a math professor at Florida State University considered to be the greatest human draughts (checkers) player who ever lived. In his forty-five-year career he lost only seven games, including the one to Chinook.

Word games were next on the list of computer achievements. In 2007 in Canada, world champion Scrabbleist David Boys fell to a computer with the unfortunate name of Quackle. In a fit of sour grapes afterwards Boys said that losing to a machine was still better than being a machine. But for how long?

Fast-forward to 2011. A monstrously fast and powerful computer called Watson, built by IBM engineers, was one of the contestants in a special edition of America's favorite quiz show *Jeopardy!* A staple on US prime-time TV since the mid-1960s, *Jeopardy!* is unusual in that the clues are in the form of answers, and contestants have to phrase their responses as questions. For example, if the question master said: "This number, one of the first twenty, uses only one vowel (four times!)", the correct response would be "What is seventeen?" To have any chance of winning, contestants need a good command of their native tongue, an impressive store of knowledge on all manner of subjects, from arts and sciences to popular culture, and to be quick on the draw.

Named not for Sherlock Holmes's famous sidekick, but for Thomas J. Watson, IBM's founder, Watson is a special-purpose supercomputer with eye-wateringly impressive hardware, including 2,800 processors, each of which would leave any high-end laptop for dead, ninety lightning-fast servers, and sixteen terabytes of RAM, connected to a database containing the equivalent of about a million books. Its power consumption is equally outrageous: 80,000 watts compared with the humble twenty-watt power demands of the humble human brain.

But the brain is hardwired for language, the product of many millions of years of evolution. It's also massively complex and interconnected, fine-tuned for the type of linguistic and knowledge-based gymnastics that makes for success at *Jeopardy!* Could Watson, a mere mass of inorganic circuitry, compete with the best human players? To find out, Watson was pitted against two past champions of the game: Ken Jennings, of Seattle, who holds the record for the most games won on the trot (seventy-four) and Brad Rutter, of Los Angeles, who racked up the highest ever cumulative winnings ($3,255,102).

Watson was represented in the studio by a tablet-like avatar, disturbingly reminiscent of the black monoliths from *2001: A Space Odyssey*. It bore IBM's blue "smart planet" logo – the temptation mercifully having been resisted to have a HAL-like eye staring out from the dark rectangle. Wired to this was a plunger connected to a buzzer, providing a close approximation to a human thumb poised at the ready. At the same moment that the question master began to read out a clue to the human contestants, the digital equivalent was fed to Watson's processors, located in a different building.

The contest was played out over two evenings, although by the end of the first the writing was on the wall. "I for one welcome our new computer overlords," Jennings quipped after his correct final *Jeopardy!* answer, prompting laughter from the studio audience. The final earnings tally was Watson $77,147, Jennings, $24,000, and Rutter $21,600.

Of course, Watson has limitations. Unlike its biological opponents, it offered nothing in the way of humor or repartee. It would be useless at any other kind of game or activity – a total failure, for example, at a hot dog eating contest or tiddlywinks – and it's incapable of emotion or feelings of any kind (or so we suppose).

It's easy to say that Watson, or any other computer, is limited by its programming. But so, to a large extent, are we. Our brains are, for better or worse, programmed by our parents and others, and the culture and events that surround us, during childhood. We think we have free will, but that could be an illusion. Each one of us knows we're conscious but we can only make assumptions about the presence or absence of consciousness in other animals and in the machines we build. At some level of complexity, consciousness and even self-awareness might emerge in computers without any active involvement from us. Then a whole new set of questions would come into play.

Just as we think nothing of turning off computers or getting rid of them altogether when we're done with them, it could be that future computers and robots, vastly more intelligent than us, would think just as little of human life. The silky voiced Watson may appear innocent enough but who's to say that a real-life HAL or Terminator won't be among its descendants a few decades from now? Trivia overlords today, masters of the planet tomorrow maybe.

Or maybe not. In science fiction a standard plot line is that an innocent-sounding, sentient computer goes off the rails and ends up threatening its human masters. Watson's great strength, though, is not plotting world domination but interpreting human language at a whole new level. That suggests its offspring will be more like the talking computer in *Star Trek*, able to understand and respond to any question in the same way as a human expert. Such machines would be invaluable in healthcare, in education, and for dispensing information generally. Miniature Watsons, more portable and less power-hungry than the original, might become commercially available in time.

Just one word of caution to any potential buyers. During a dry run for the *Jeopardy!* challenge, when the computer was finally out-buzzed in coming up with an answer, Watson's avatar briefly flashed an angry orange. It happened so quickly that few on the set seemed to notice it, but Rutter caught the moment and later joked about it, being, as he put it, "afraid of Watson's progeny coming back from the future to kill me".

MIND MELD

Terminator-like beings will hopefully always remain in the land of fiction. In fact, far from manifesting itself as the metallic monsters of movies, advanced AI may creep up on us from an entirely different direction. Rather than the creation of an entirely separate intelligence, hardwired in silicon, or in the form of a computer programmed to act as smartly as a person, it seems more likely that we'll first see the emergence of super-intelligence in systems in which brains and computers are linked together.

To some extent such systems already exist. Whenever you use a laptop or even a games machine, like a Nintendo DS, your brain and the computer's processor are working in tandem – the one depending on the other in a rapid pas de deux of call and response. It's just that the connection isn't very direct or intimate.

In the early days of personal computing the only way to get data manually into a computer was by throwing switches on a control panel (in the case of the Altair 8800) or a standard QWERTY keyboard. Today, there are mice, light-pens, touchpads, touchscreens, voice recognition systems, and motion sensors, which enable us to interact more intuitively and in a far more immediate, information-rich way than by tapping alphanumeric keys. The controller-free motion sensor arrangement of the Xbox Kinect is a dramatic example of how anyone, devoid of even the slightest training or tech-savviness, can now interface effortlessly and instantly with a complex box of circuitry. Most of the time we're still hampered by the klutzy keyboard. But waiting in the wings is the ultimate link between man and machine: the brain-computer interface in which there's a direct communication pathway between electrical activity in the brain and the computer's hardware.

To date, most research in this direction has been done on applications of neuroprosthetics aimed at restoring impaired hearing, sight, and movement to those who are handicapped. Thanks to the amazing plasticity of the brain's cortex, signals from implanted prostheses can, after a period of adaptation, be handled by the brain as if they were coming from the body's own sensors or effectors. Tens of thousands of people worldwide have been give cochlear implants, and systems aimed at restoring sight, including retinal implants, are starting to move out of the prototype stage.

Also on trial are a number of motor neuroprosthetics for giving back movement to people who've been paralyzed, or helping them control computers or robotic arms. In 2005 quadriplegic Matt Nagle became the first person to control an artificial hand using a brain-computer interface. Implanted in Nagle's right precentral gyrus (an area of the brain responsible for arm movement), the ninety-six-electrode BrainGate device, supplied by Cyberkinetics Neurotechology, allowed Nagle to control a robotic arm, as well as a computer cursor, lights, and TV, by thought alone.

So-called invasive brain-computer interfaces are the most effective form of connection. Because they involve hardware that's implanted directly into the substance of the brain during neurosurgery, they're able to generate the highest-quality signals. One disadvantage of them is that the brain, like other parts of the body, reacts to foreign objects buried in it by producing scar tissue which can degrade the signals over time.

Partially invasive devices, placed just inside the skull, get around the scar tissue problem at the cost of reduced signal resolution. In 2004 Eric Leuthardt and Daniel Moran from Washington University in St Louis trialed a system in which electrodes, embedded in thin plastic pads, were placed beneath the dura mater (the protective, tough membranous covering of the brain), just above the cortex, to measure the electrical activity of the brain. In a later trial, the researchers enabled a disabled teenage boy to play Space Invaders using this kind of implant.[30]

Non-invasive brain-computer interfaces are the kind everyone would prefer, but they suffer from relatively poor signal resolution. The five- to eight-millimeter thickness of the skull is enough to dampen signals, dispersing and blurring electromagnetic waves created by the brain's neurons. Still, it's surprising how effective even a few strategically placed electrodes around the head can be. In the game of Mindball, developed by the Swedish company Interactive Productline, players attempt to push a ball across a table by becoming more relaxed and focused. It can literally be child's

Figure 11 Players compete at Mindball at the Dundee Science Centre. Credit: Bob Douglas.

play – a five-year-old can do it – with the help of a fabric band around the head containing electrodes which monitor the type of brain waves being produced. Two players compete by trying to be the first to get their ball across a goal-line, the winner being he or she who can enter the most relaxed state, characterized by a preponderance of theta and alpha waves.

Mindball is a novel extension of biofeedback devices that were all the rage back in the 1970s. It's a far cry from the technology used by Clint Eastwood's character in the 1982 film *Firefox*, which enables a fighter plane and its weapons to be controlled by thought alone. But a step toward genuine thought control is now being offered by the San Francisco-based company Emotiv (corporate motto: "You think, therefore, you can") in the form of its $300 EPOC "neuroheadset". This is being marketed as both a games control device and as a potential aid for the disabled. Its sixteen electrodes monitor brain activity related to facial movements, in order to infer the emotional state and intentions of the wearer. The electronic data are then fed to software in order to control various applications, from playing games to steering a wheelchair.

At the Free University of Berlin, Raul Rojas and his team have adapted Emotiv's product to enable a car to be guided by thought alone. The experimental car – a modified Volkswagen Passat – is also kitted out with laser radars, microwave radars, and stereo cameras so that it can detect obstacles in a full 360-degree sweep and sense other vehicles up to 200 meters away. The driver can take the wheel, as it were, whenever he or she wants to, by mentally instructing a cube displayed on a screen to move right or left. At other times, aided by its radars and cameras, the car can steer itself or take navigational cues from a variety of other control devices such as an iPad or iPhone. The ultimate goal of the research is to develop a roadworthy semi-autonomous vehicle that, when necessary, can be told what to do via a neural interface.

Canadian company InteraXon is also getting in on the neural interface act, wowing crowds at the 2010 Winter Olympics in Vancouver by enabling attendees there to control lights on the CN Tower, 3,350 kilometers away, in Toronto, simply by thinking about it. But a greater challenge lies ahead. It's to go beyond controlling devices by general changes in brain activity, such as those which accompany relaxation, to specific commands and even, eventually, to communicate with a computer through words that aren't vocalized but merely formed in our minds.

Inevitably, there are going to be limits beyond which non-invasive brain-computer interfaces can't be pushed. The bony casing of the skull may always prove too much of a barrier to allow individual thoughts or intentions to be picked up by EEG-like contacts on the surface. We'd probably baulk today at the idea of undergoing surgery to have implants placed beneath the skull just so that we can talk more easily with our PCs or Macs. But the day may come when invasive brain-computer technology becomes so effective and, frankly, indispensable, that we overcome any such squeamishness. Some of the technology that may be involved has already been tested on rodents. Nanowires, which might one day snake through the brain's capillaries, sending and receiving information, have been shown to grow chemically in rats' brains. Meanwhile in the nascent field of optogenetics, tiny, precision-guided laser beams have been used to trigger specific memories and behaviors in rats, raising the prospect that they might also eventually be used to activate and deactivate individual neurons in a human brain, in the same way that transistors are switched on and off inside the circuits of a computer.

Step by step, if current trends continue, the relatively crude thought-control devices available today will become more

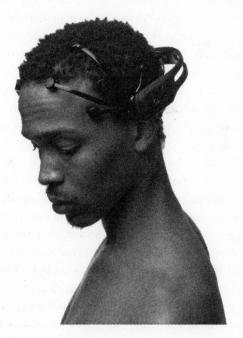

Figure 12   The Emotiv EPOC neural headset. Credit: Emotiv.

sophisticated and widespread, until people everywhere will be able to interface neurally with the World Wide Web. Simply framing a question in your mind will then produce an answer drawn either from your own collection of neurons or, almost without realizing the difference, from the vastly greater database of knowledge on the Internet. Social networking and computer-based communications like Skype will evolve into synthetic telepathy, because when all our minds are connected to the Internet we'll effectively be in telepathic communication with each other. The boundaries between brains and computers, and one brain and another, will begin to break down so that effectively, over time, we'll become part of a growing planetary mind.

At which point it could all go spectacularly pear-shaped.

HOMO CYBORG

The idea that we'd volunteer to have devices stuck inside our heads so that we can be mentally linked to the global information system may seem absurd and ridiculously far-fetched. But everything is trending that way, and at a remarkably rapid pace. We're spending more and more time on the Internet, relying on it increasingly for work and play. Anything that would make it easier to connect to this marvelous, all-pervasive resource without the need for clumsy keyboards or even more natural systems such as voice recognition, is likely to be adopted. As for the invasiveness of the technology, well, that may not be such an issue in time. With the dawn of nanotechnology may come interface equipment of microscopic dimensions, which can be inserted with no more discomfort than LASIK eye surgery or a small tattoo, and upgraded as easily as buying a new SIM card for a mobile phone.

Eventually, it may become socially acceptable, and even the norm, for babies in developed countries (in other words, those whose parents can afford it) to have neural implants fitted shortly after birth so that their brains can begin adapting without delay to being part of a communal consciousness. That idea might seem repulsive today. But in the years ahead it might seem equally unacceptable to allow a child to grow up isolated from the rest of the nurturing Earth-brain and be deprived of all the advantages it will have to offer.

Images spring to mind of *Star Trek*'s most terrifying villains, the Borg. They, too, are given cybernetic implants (against their will) in order to be connected to a "hive" mind. Will the future Internet, with its multitudinous brain-computer connections, turn us all into drones? When Captain Picard is assimilated to become Locutus of Borg he utters the chilling words: "Resistance is futile. Your life as it has been is over. From this time forward, you will service us." Our natural inclination may be to fear that by plugging ourselves into some giant collective we'll lose our self-identity and become swallowed up by a monstrous man-machine assemblage – call it Homo cyborg.

Self-identity is a funny business once we start interacting with each other via computers. We don't have to let people know what we're "really" like. We can adopt avatars when we play on-line games. Without the need for face-to-face conversation we're free to create new personas, becoming confident and outgoing, brave and beautiful, whereas in real life we might be more shy and mundane. Our existence in the digital landscape can be wildly different from who we are in the flesh and in our humdrum physical world. In any event there's no doubt, given the runaway success of social networking systems such as Facebook, that we relish connecting with one another in cyberspace, at least as much as we do in real space. We're happy, it seems, to become part of a larger whole, as if we were individual cells in a giant electronic brain.

Even a few years ago it would have seemed odd to hear someone apparently talking to themselves in the street or on a train. But now to see people with mobile phones virtually glued to their ears has become the norm. Hands-free sets are increasingly common. The fact is, we like being joined to an electronic community – being able to be in touch with others, anywhere on the planet, almost instantaneously. If there were a genuine planetary mind now with readily available and easily implantable neural interfaces, it's hard to imagine there wouldn't be a vast clamoring around the world to be part of it.

But computers can be infected with viruses, worms, and Trojans. If computers are intimately connected to our brains, the horrible prospect arises of similar digital infections getting inside our heads. Equally appalling is the possibility that malicious agents, for example authoritarian regimes, might seek to alter people's minds via brain-computer links. Propaganda could be delivered direct to our neurons, without us realizing it. Thoughts could pop into our heads, which we might even mistake for our own, aimed at controlling or manipulating our behavior. Clearly, before we take the plunge of wedding ourselves to any future Earth-brain there'll need to be far more effective firewalls, virus and malware detectors, and other safeguards than are available to protect our humble PCs, which regularly fall victim to these unwelcome intruders. If eventually the entire human population is linked to a planetary mind, a single nasty bit of computer code unleashed by some teenage malcontent could crash our entire species or turn us into zombies.

On the other hand, the great communal mind of the future might offer us a form of immortality. Today, the best we can do in the way of a digital legacy is to leave words and pictures online for others to see, for example via Facebook. Several companies are taking the first steps towards enabling people to make robo-clones of themselves, an initial goal being to enable someone to create a lifelike digital representation of him- or herself that can continue long after the biological body has decomposed. But when a great deal more is known about how the brain works and is arranged, it may become possible to replicate its structure artificially so that we can effectively upload our consciousness to the Earth-brain, thereby escaping the inevitable decay of our organic frames. What then of the old human race? We may all end up as disembodied minds inside some artificial matrix – sparks of awareness that are more machine-like in nature than anything we'd feel remotely comfortable with today.

Whether this mental merging with technology will prove good or bad for us, or even be an existential threat, is impossible to say at present, but it seems likely that in some guise or other it will happen and that it will be massively transformative. It may be one aspect of what's been called the singularity – a phase of exponential development when technological change becomes so fast that in our current state we can't meaningfully speculate where it will lead.

TOTAL IMMERSION

In the more immediate future, given the way events are unfolding, it's clear that people are going to become more and more used to being immersed as watchers or participants in vivid artificial landscapes. Three-dimensional films and television are becoming commonplace. Simulators, for serious applications such as pilot training or just for pleasure, have been around for years. Second Life is among the more popular and impressive virtual worlds now available on the Internet. Like World of Warcraft and other such sight-and-sound enriched multiplayer games, it allows the player to create a character from scratch and interact with other avatars in a cosmos of the imagination. Every month, the number of people engaged in virtual worlds is increasing by about fifteen percent.

So-called virtual reality uses additional gadgetry, such as headsets and "data-gloves", to more fully immerse a person, engaging the senses to a much greater extent with the illusion. Even so, it's easy to tell virtual reality from actual reality; we're never in any doubt about the nature of what we experience. But that may change. Beyond virtual reality lies what's known as simulated reality, an experience that would be hard or impossible to distinguish from the physical world. *Star Trek*'s holodeck is a fictional example. It isn't too hard to imagine that in a few decades' time something along the lines of holodecks will be available in public gaming arcades or in the homes of those who can afford them. By then, systems like Wii and Kinect will be regarded as museum pieces.

The illusion of an alternative reality available in holo-rooms will be so convincing and overpowering that it may also become seriously addictive. When any experience imaginable is available on demand in these theatres of unreality, how will we cope with emerging back into the comparative dullness of everyday life?

As computer worlds become increasingly immersive, we're heading for a situation in which emotional and psychological problems associated with them intensify. Consider the experience of a single nurse, Linda, which was covered by CBS Up to the Minute News in 2008. Linda entered Second Life as an attractive, red-haired character called Cala. In this on-line world, Cala met the man of her dreams, Richie, and after a four-month courtship they became engaged. In the interview with CBS, Linda remarked that when Richie proposed to Cala it felt to her (Linda) like a real life proposal. Following a fantasy wedding complete with minister, bridesmaids, groomsmen, cake, reception, and honeymoon, Richie and Cala were married. Cala and Richie's flesh-and-blood counterparts both admitted they'd actually fallen in love. Unfortunately, he was already married and his real-life wife knew nothing of his Second Life shenanigans. Bizarrely, although Richie's human creator didn't want to be interviewed (for obvious reasons), he did agree to talk to the CBS reporter as his Second Life character. The reporter duly created an avatar for herself and entered the virtual world to meet Richie at the Second Life home he shared with Cala. Asked if he thought he was cheating on his wife, his response was, "I suppose it's cheating in a way, but it's a meeting of minds not bodies." He also admitted that his Second Life marriage was stronger than his real-life marriage.

If such situations are already arising in a system as relatively crude as Second Life it raises serious questions about the societal and psychological effects that future simulated realities will have upon us. There's also the prospect, mentioned earlier, of actual conscious entities inhabiting these digital worlds – either artificially created ones or uploads of human minds. It seems likely it'll become increasingly difficult to distinguish between relationships in the natural world and the anything-goes universe of the communal Earth-brain. At the very least, there'll be a market for psychiatrists specializing in simulated reality addiction and associated personal problems.

THE WORLD THAT NEVER WAS

More disturbingly, the possibility has been suggested that we may already be living in a simulated reality. What we take to be the natural cosmos may be just a simulation set up by some intelligence beyond our ken. According to this idea, we're no more than self-aware characters in a holodeck novel.

In 1988, the Austrian computer scientist Hans Moravec became the first to suggest what's become known as the simulation argument. Ten years later, the Swedish philosopher Nick Bostrum, at the University of Oxford, explored the ramifications of this argument in detail.[31] Bostrum started from the premise that it's not unreasonable to suppose that an advanced civilization would eventually develop the means to create a simulated reality – and not just one simulated reality but very likely a huge number of them, for research purposes or just for fun (maybe in the form of a commercially available game). Now, Bostrum asked, what's more likely: that we're the first species who in time will do this, or that we're actually the characters in one of the countless simulations created by other intelligences? Remember, if we are just conscious characters in a simulated reality we wouldn't know it because to us it would seem like the "real world".

Maybe the virtual characters in a simulated reality can become so advanced that they can fashion their own simulated realities, and so it might go on, layer upon layer. Following this line of reasoning, we may not even be in a "first generation" simulated reality but one that's nested much deeper. Suddenly the multiverse concept of the cosmologist starts to seem small beer by comparison. Reality itself might be a relative concept and in truth there's an untold number of simulated realities, many of them inhabited by sentient beings who are under the illusion that they live in the one true reality, arranged in a mind-bogglingly vast hierarchy.

Unfortunately, simulations can be ended with disturbing ease by their creators. Just as we can close down a computer game with a simple click when we've had enough of it, so, if we were virtual characters in some much more elaborate charade our existence could be rudely curtailed by a higher being getting bored of the goings-on in our own little cosmos. Press "quit" and we and our fictional universe are returned to the digital recycle bin of a computer in a more real level of reality.

Perhaps if we live in a simulated reality we'll find out about it, just as some of the protagonists in *The Matrix* do. Then we might find some way to pop out of our simulated reality into the reality level above us. After all, whether we're participants in a simulated reality or not, we are conscious entities and we're ambitious and smart.

On the other hand perhaps in this case ignorance is a blessing. The obsolescence of humanity would be tragic enough. The discovery that we'd been totally irrelevant all along would be unendurable.

SURVIVAL TIPS

Internet and gaming addiction is probably the worst that anyone can suffer because of computer technology today. Avoiding more severe future problems, like having your mind scrambled by a virus that crosses a neural interface, may call for more caution and multiple levels of software protection. If ever a large fraction of the human population becomes directly connected with the World Wide Web, governments, companies, and individuals will have to think carefully about how to protect the whole system – silicon-based and carbon-based – from being knocked out by accident or intent.

If computers and robots become intelligent enough on their own to rival or surpass human intellect, perhaps they'll have to include morality circuits, or at least have Asimov's Laws of Robotics hardwired into them.

CHAPTER 5

KABOOM!

We spend most of our lives in a comfort zone between fire and ice. In terms of its orbit, the Earth is ideally placed between Venus, which is unbearably hot, and Mars, which, for the most part, is extremely cold. Our planet's average surface temperature is a benign 15°C (60°F). As a rule we pernickety humans tend to avoid local hotspots, such as erupting volcanoes, and the more or less permanently frozen poles.

But sometimes nature forces extremes of temperature upon us, along with other environmental hardships. The effects may be local, regional, or global in scope. When the Earth erupts in a molten fury or freezes over in a new ice age or worse, the continuity of life, and of human life in particular, may hang in the balance. Oddly enough, a massive eruption, which is fantastically hot if you're unlucky enough to be nearby, can be one of the triggers for a stint of extreme worldwide cold.

BLASTS FROM THE PAST

One of the most eye-catching volcanic eruptions in modern times was that of Mount St Helens, a lofty peak lying 150 kilometers south of Seattle, in Washington state. At 8:32 AM on Sunday, May 18, 1980, about three cubic kilometers of the mountain came crashing down in a massive rock debris avalanche triggered by an earthquake measuring 5.1 on the Richter scale. Nearly 600 square kilometers of forest was flattened by the fiery blast wave or buried beneath volcanic deposits, fifty-seven people were killed, and 300 kilometers of highway were destroyed. At the same time, a forbidding mushroom cloud of ash rose thousands of meters into the atmosphere and drifted downwind, darkening the sky and raining thick, grey ash over eastern Washington and beyond. The eruption lasted nine hours, but Mount St Helens and the

Figure 13 The eruption of Mount St Helens. Credit: USGS Cascades Volcano Observatory.

surrounding terrain were altered beyond recognition within moments.

An eruption of a very different kind has been going on, in fits and starts, in Hawaii since January 1983. For more than a quarter of a century, Kilauea has been erupting – non-violently, but ejecting in total more material than Mount St Helens managed in its sudden spectacular outburst. A new vent opened at Kilauea as recently as March 2011, sending lava shooting twenty meters into the air.

Both Kilauea and Mount St Helens are dwarfed, however, in the amount of lava, rock fragments, and dust erupted, by two other volcanic extravaganzas in the twentieth century. These were the eruptions of Pinatubo in the Philippines in 1991, and Novarupta in the Alaska Peninsula in 1912, which threw out an estimated ten and twelve cubic kilometers of material, respectively. Along with the ten billion metric tons of magma which spewed out of Pinatubo were twenty million metric tons of sulfur dioxide. This gas was oxidized in the atmosphere and formed a global haze of sulfates and sulfuric acid, the effect of which was to cut the amount of solar radiation reaching the Earth's surface by about ten percent and reduce temperatures worldwide by about half a degree Celsius. At the same time, ozone levels at mid-latitudes reached their lowest recorded levels, while in the southern hemisphere winter of 1992, the ozone hole over Antarctica widened to its largest ever size up to that point.

But even Pinatubo and Novarupta are small fry compared with other volcanic monsters of the past. In 1883 the island of Krakatoa in Indonesia blew apart, killing at least 40,000 people and throwing out twenty cubic kilometers of rock, ash, and pumice, in an ear-splitting explosion that was heard 3,500 kilometers away in Perth, Australia, where it was mistaken for cannon-fire. Barograph readings showed that the pressure wave from the event circled the Earth seven times before finally fading out. The settlement of Merak on the northwestern tip of Java was washed away in seconds by a tsunami towering almost fifty meters high, and ships as far away as South Africa were rocked as the residues of powerful tsunamis buffeted their hulls.

More violent still was the destruction of Santorini, a small island in the Aegean, 3,600 years ago. A giant central lagoon is all that remains of the great volcano that erupted in about 1620 BCE spewing vast clouds of dust and ash, and generating a huge tsunami that inundated the nearby island of Crete. This is the event, according to some historians, which triggered the demise of one of the world's great early civilizations – the Minoans. The blow-up of Santorini, it's also been theorized, may be the historical counterpart to the legend of Atlantis. At least thirty cubic kilometers of magma, rock bombs, dust, and other debris poured out of Santorini, ranking the eruption among the top seven or eight biggest of the past 10,000 years, and making it perhaps the most influential in terms of its effects on the course of western history.

A MEASURE OF DESTRUCTION

It's hard to imagine that there could be volcanic eruptions which dwarfed those of Krakatoa and Santorini. But there were – and there will be again. One way volcanologists classify eruptions is in terms of the Volcanic Explosivity Index (VEI). Devised in 1982 by Chris Newhall of the US Geological Survey and Stephen Self of the University of Hawaii, it gauges the relative violence of volcanic blow-outs. Factors that go into determining the VEI are the volume of material ejected, the height of the cloud of debris from the eruption, and qualitative observations about the nature of the explosion. The VEI range goes from 0, for a non-explosive, gentle burbling or spraying of lava, to 8 in the case of the most prodigious, mega-cataclysmic supervolcanoes.

None of the volcanoes we've talked about so far are supervolcanoes. Kilauea, for example, despite the volume of stuff it's put out over the years, is a very mild erupter – mild enough for hoards of tourists to visit it safely each year (except during times when it gets a little feisty and starts to whoosh lava a few tens of meters into the air). It has currently a lowly VEI of 1. Famous Mount Etna, on the east coast of Sicily, received a rating of 3 for its 2002–2003 efforts, while the 2010 air traffic-disrupting effusions of Iceland's Eyjafjallajökull (every newsreader's nightmare) went one better with a 4. With an index of 5 comes the Mount St Helens eruption, followed in category 6 by Pinatubo, Krakatoa, and Novarupta.

Santorini, for its circa 1620 BCE detonation, earns an impressive VEI of 7. Matching this score is the little-known 186 CE eruption of Taupo on New Zealand's North Island. Lake Taupo, the second-largest freshwater lake in Oceania, is actually the submerged caldera of a colossal volcano that had its last major eruption over 1,800 years ago. Only one explosion in recent history has made it to 7 on the Volcanic Explosivity Index. This is Mount Tambora, on the island of Sumbawa in Indonesia. Tambora began to rumble ominously in 1812, and reached a crescendo with a mind-numbing eruption in April 1815. Roughly 160 cubic kilometers of ejecta issued from Tambora's 1815 outburst, making it the largest volcanic eruption in recorded history. The vast quantities of dust and ash entering the atmosphere lowered temperatures worldwide for months afterwards, and the following year – 1816 – became known as the "year without a summer". But what we know about Tambora and its effects was learned in retrospect. In fact, astonishingly, it took 160 years for the truth to come to light. Examining ash layers found in Greenland ice cores, scientists were amazed to discover the extent of the Tambora phenomenon – an eruption ten times more powerful than any other in the past two centuries, including the awesome Krakatoa.

Tambora, just east of the island paradises of Bali and Lombok, once rose to a height of over 4,000 meters. That was before its spectacular awakening after a long period of dormancy lasting five millennia. For three years, beginning in 1812, Tambora spewed steam and ash, and was rocked almost incessantly by earthquakes. Unbeknown to anyone at the time, seawater had found its way through cracks in the rock into the magma chamber deep under the volcano. Disaster was now inevitable. Gradually the pressure of superheated steam, in the heart of the mountain, built up and up until it could no longer be contained. On April 10, 1815, Tambora was blown apart in a series of devastating explosions. Volcanic ash and gas shot high into the stratosphere, while at ground level that deadliest of all volcanic phenomena, a pyroclastic flow – a fast-moving flow of superheated rock and ash – raced across the surrounding landscape at speeds that would have overtaken a Ferrari, incinerating and burying everything in its path. About 10,000 people died immediately. When the pyroclastic flow reached the ocean, it set off tsunamis which ravaged the populations on nearby islands. Ash from the eruption landed as far as 1,300 kilometers away. Where it fell thickest it killed crops and other plants, eventually leading to the deaths of another 80,000 to 90,000 people from famine. The total volume of material ejected was nearly 100 times that thrown out by Mount St Helens.

The vast quantities of gas and aerosols injected into the atmosphere by Tambora blocked sunlight and caused temperatures to tumble around the world, most acutely over the northern hemisphere. Crop failures became common in Europe, with hunger and disease following in their wake.

One of the first modern works of horror and science fiction came about indirectly because of the Tambora eruption. The poets Lord Byron and Percy Shelley were staying by Lake Geneva in 1816, and, as a result of the bad weather, spent much of the summer with their friends and family entertaining indoors. During one such gathering, Byron proposed that everyone present try to write a ghost story. Byron himself, inspired by the brooding skies of that gloomy season, wrote the poem "Darkness". Shelley's wife, Mary, came up with an even more menacing creation – her novel *Frankenstein*.

WHEN ALL HELL BREAKS LOOSE

The eruptions of Tambora, in 1815, and Taupo, New Zealand, in 186, were the two biggest of the past 10,000 years – roughly the period for which humans have been civilized. Both score 7 on the Volcanic Eruptivity Index. Both blasted out around 100 cubic kilometers of rock, dust, and ash. But neither were supervolcanoes. That term, first used in a BBC *Horizon* program in 2000, is generally reserved for eruptions involving at least *1,000* cubic kilometers of ejecta – ten times the size of Tambora and Taupo, and similar in destructive capacity to a one-kilometer-wide asteroid barreling into the Earth. Supervolcanoes have a VEI of 8, the maximum value recognized. According to the geological record, one of these monsters explodes on average every 100,000 years (making it ten times more likely than an asteroid impact capable of similar devastation). The last happened about 74,000 years ago on the island of Sumatra in western Indonesia. Known as the Toba super-eruption, its site is marked today by Lake Toba – at 100 kilometers long by thirty kilometers wide, the largest volcanic lake in the world.

Even by supervolcano standards Toba was impressive, unrivaled by any other eruption over the past 25 million years. The amount of molten rock and other stuff it poured out came to around 2,800 cubic kilometers – more than double the volume of Mount Everest. Imagine 10,000 Mount St Helens going off in the same place at the same time and you get the idea. Toba occurred at a time when Neanderthals and more modern humans coexisted in Europe and much of Asia, and its after-effects may have brought immense hardship to our ancestors – perhaps, according to some suggestions, even pushing them to the brink of extinction.

Immediately following the blast, global temperatures fell – by how much and for how long is a matter of debate. A key factor is the amount of sulfur dioxide and aerosols released during the eruption and what happened to them once they got high into the atmosphere. Different theoretical models produce different results. What seems likely is that the Toba event put significant stresses on humans and their close relatives who were around at the time.

It's generally accepted that the Toba super-eruption caused a worldwide slump in average temperatures of between 3 and 5°C, and as much as 15° at higher latitudes. The ash fall-out was prodigious. A blanket of ash at least fifteen centimeters deep covered all of South Asia, and in places the deposition was much greater – six meters (twenty feet) at one site in central India, and nine meters (thirty feet) in parts of Malaysia. Flora and fauna alike in southeast Asia must have been devastated, with very few plants surviving, and there may have been a planet-wide die-off.

Certainly, the Toba blow-out happened slap bang in the middle of a period, between 100,000 and 50,000 years ago, when the human population plummeted. This has led to the Toba catastrophe theory, according to which the effects of the eruption were so severe that the global population of *Homo sapiens* was slashed to 10,000 individuals or less. The catastrophe theory, although controversial, is backed up by genetic evidence, which suggests that all of us alive today, despite our apparent diversity, are descended from a very small group of people, perhaps as few as 1,000 breeding pairs, about 70,000 years ago. The environmental pressures on this perilously tiny group of human survivors could only have been made worse as a result of the Toba super-eruption.

Somehow our species clawed its way back from its low-population crisis, and not only *Homo sapiens*, but also, for a while, the Neanderthals in Europe and, it seems, a small-brained relative of ours, *Homo floresiensis* (popularly known as Hobbits). The Toba disaster, the suggestion goes, would have forced a lifestyle change on humans at the time. The die-off of vegetation, and the cooling and drying of the climate resulting from the volcanic fall-out, might have altered the migratory habits of our ancestors and compelled them to adopt new and ingenious methods to gain access to whatever scarce food sources were available. As separate little colonies developed independently, they would have sown the seeds for the differences between races seen today. Ultimately, our species may have benefited from the ordeal of Toba; it made us tougher, smarter, and more reliant on our wits and on our latent talents for communication and cooperation. The Neanderthals evidently didn't fare so well in the final analysis. But we came desperately close to annihilation in those troubled times.

DANGER FROM THE DEPTHS

Earlier super-eruptions have been linked to mass extinctions, when whole swathes of life, animals and plants, were erased within a short space of time by geological standards. The Permian mass extinction of 250 million years ago, which wiped out more than ninety percent of the animal species on Earth, is thought to have been tied to a colossal eruption event associated with the Siberian Traps. These Traps (the name comes from the Dutch word for "stairs") make up a vast region of igneous rock which formed from the outpouring of between one and four million cubic kilometers of lava, and today covers a large part of Siberia. The activity took place over about a million years and involved individual eruptions each of which may have rivaled the Toba eruption. So dramatic were the effects on the global ecosystem that it took land life about thirty million years to recover.

The Siberian Traps are just one example of what are called large igneous provinces (LIPs), vast outpourings of lava caused when giant blobs of magma in the Earth's mantle burp their way to the surface. Another such catastrophic gushing of lava from the bowels of the planet gave rise to the Deccan Traps, which at the time they formed, between sixty and sixty-eight million years ago, buried most of India under molten rock. The peak of the eruption has been dated to sixty-six million years ago, just prior to the mass extinction at the end of the Cretaceous Period when the last of the dinosaurs and many other animals and plants disappeared from the fossil record. It's widely accepted that the dinosaurs were finished off mainly by a large asteroid collision, as described in chapter 7. But environmental fall-out from the Deccan Traps could also have contributed to their downfall.

The source of the underground magma that occasionally rips through the surface to form LIPs is a matter of intense study. Conventional wisdom has it that LIPs come from relatively young molten rock in the upper mantle formed when oceanic crust dives under continental crust in a process called subduction. But the latest theory in town, still highly controversial, is that the enormous eruptions of lava that coincided with some past mass extinctions are linked to two very hot and deep-lying blobs of mantle material which have existed almost as long as the Earth itself.[32] Seismic studies have revealed two unusual regions some 2,800 kilometers down, beneath Africa and the Pacific Ocean, and a team from the University of Oslo, Norway, recently showed that most LIPs formed while one of these two regions lay directly beneath that part of the surface.[33]

THE YELLOWSTONE MENACE

There's no doubt that the Earth will experience more super-eruptions. Some of the supervolcanoes that have erupted in the past and caused such global mayhem retain the capacity to do so again. Possible contenders for the next super-eruption include the Phlegrean Fields volcano west of Naples, Italy, and Lake Taupo in New Zealand, as well as locations in Indonesia, the Philippines, Central America, Japan, and the Kamchatka Peninsula in eastern Russia. But of all the candidates none has been so talked about, or instilled such popular fear, as one of the most visited, picturesque, and unique areas of the planet, right in the heart of the United States: the Yellowstone caldera in the northwest corner of Wyoming.

The Yellowstone caldera, occupying about half of Yellowstone National Park, measures some seventy-two kilometers (forty-five miles) by fifty-five kilometers (thirty-four miles) and is the site of numerous past eruptions, many of them in the range of ordinary volcanoes but a few in the supervolcano class. The most recent of these gargantuan outbursts, about 640,000 years ago, is thought to have been responsible for the demise of many of the larger mammals in North America at the time, including camels, rhinos, and elephants, which choked on toxic gases or starved to death following the die-back of vegetation blanketed by the continent-wide ash cloud from the event. Other super-eruptions of the Yellowstone supervolcano happened 1.3 million and 2.1 million years ago. The average period between eruptions, geologists have determined, is about 600,000 years, which means we're due for another one at any time.

This wouldn't be good news for America. Blanketed in a layer of ash a meter thick (and a great deal more in areas close to the eruption), the country would be rendered virtually unfit for human habitation. Mid-western states, home to much of the nation's food production and industry, would be hit especially hard. But the effects would be felt much wider afield. As in the case of the Toba eruption, global cooling, the mass dying of plants, and then the mass dying of animals and people would follow in the days, months, and years after the cataclysm. In 2005, a working group of the Geological Society of London assessed the consequences of a contemporary super-eruption. "[P]ronounced deterioration of global climate would be expected for a few years following the eruption," it said. "Such events could result in the ruin of world agriculture, severe disruption of food supplies, and mass starvation. The effects could be sufficiently severe to threaten the fabric of civilisation."[34]

Not surprisingly, having seen dramatizations on TV of what a super-eruption can do, people get a little jittery when reports come through of fresh activity in the nation's favorite national park. "Run for your lives ...Yellowstone's going to explode!" read a (slightly tongue-in-cheek) news headline by the Associated Press on January 10, 2009. Hundreds of minor earthquakes in the proceeding weeks had set nerves jangling, reminding everyone that the beautiful landscape and photogenic curiosities, such as geysers, mud-pots, and hot springs, in this corner of Wyoming sit atop a supervolcano that *will* burst into life again, sooner or later.

Small earthquakes aren't unusual in Yellowstone, and the signs of geothermal activity are everywhere to be seen, from the predictable appearances of the Old Faithful geyser to the constantly bubbling, burbling, sulfurous cauldrons of hot water and mud to be found all over the Park. These crowd-pleasing features are just mild expressions of the colossal forces that are steadily building up below.

Between six and sixteen kilometers beneath the picture-perfect scenery of Yellowstone is a giant magma chamber, which is slowly but surely filling with molten rock from the underlying mantle. It's an estimated fifty kilometers long, thirty kilometers wide, and ten kilometers deep, and is fed by a magma plume that rises at a 60° angle from at least 660 kilometers beneath the Earth's surface. The deepest part of the plume lies under the town of Wisdom, Montana, about 240 kilometers from Yellowstone National Park. Trapped gases are steadily increasing the pressure inside the magma, and although some of that pressure is gently relieved on a daily basis by the various geothermal features that attract visitors to the park, it isn't enough. There may be minor eruptions first: no one can be sure that the next blow-out will be a monster one. But, at some point in the future, the pressure inside the subterranean chamber will reach a critical level, the overlying rock will be split apart, and the gas-laden magma will erupt explosively over a wide area at the surface. On that fateful day, more than 1,000 cubic kilometers of magma might burst into the light of day and bring hell to North America and beyond.

SURVIVAL TIPS

The good news is that the Yellowstone caldera is one of the most closely monitored and studied volcanically active regions on Earth. Scientists measure it daily with a battery of instruments. Between 2004 and 2008 they found that the upward movement of the caldera floor was about eight centimeters per year – more than three times greater than ever observed since such measurements began in 1923. But by the start of 2010, the US Geological Survey announced that "uplift of the Yellowstone Caldera has slowed significantly" and it now continues at a reduced pace. Scientists with the USGS, University of Utah, and National Park Service have said that they "see no evidence that another … cataclysmic eruption will occur at Yellowstone in the foreseeable future."

If Yellowstone does blow in the coming years, the warning signs won't necessarily tell us if it's going to be a super-eruption or a smaller event, of which there have been many – the last just 70,000 years ago. It's also possible that we wouldn't have much of a heads-up in the case of a global-scale event. There's an argument which says that in order to get out a huge volume of magma in one go, a super-eruption might involve a different mechanism than smaller eruptions – for example, the downward extension of a fracture that taps magma deep down and brings it extremely rapidly to the surface. In such a situation, warning signs may be few and short-lived.

Even if we did know a few days, weeks, or months in advance of an impending supervolcano – in Yellowstone or anywhere else – it's not obvious how people and their governments could constructively respond. Getting out of the immediate area as fast as possible would be a given. But this would involve evacuating at least a several-hundred-kilometer-wide radius of ground zero, and the panic and traffic jams that would ensue hardly bear thinking about. Unfortunately, although it would minimize the loss of human life, evacuating an entire continent isn't really an option. For people at a safe distance from the actual lava flows, health problems from a much wider dust fall-out would be the immediate problem so that face-masks and, preferably, more heavy-duty breathing apparatus, of the kind used to protect against noxious fumes, would be the most essential survival aid. Beyond that, people would probably be advised to stay indoors as much as possible and try to have available a few weeks' supply of food and water until the worst of the eruption was over.

CHAPTER 6

FORBIDDING PLANET

A harsh winter, when temperatures barely climb above freezing for weeks on end, and snow and ice lie thick on the ground, can bring transport systems to a virtual standstill. Power and communication networks can break down or become overstretched, causing more hardship. Multiply that situation a hundred- or a thousand-fold across whole swathes of the planet and the wheels of our high-tech civilization would start to grind to a halt. If temperatures plunged low enough for long enough, down as far as the mid-latitudes, the food chain itself would be disrupted and the well-being of millions called into question.

Worldwide cool-downs are a fact of life on Earth. They can be temporary, lasting a few months or years, or protracted, going on for millennia. They can be triggered by a sudden event, like a big volcano blowing its top, or a more gradual process, involving subtle changes in the Earth's motion in space or the behavior of the Sun. Severe global freeze-ups pose the threat of a megacatastrophe. But exactly how they come about and when the next one will take place are questions that still perplex science.

FIRE AND ICE

Volcanic eruptions, as we saw in the last chapter, can affect the climate across large parts of the globe. The effects range from mild to severe, and may be short-term or long-term, depending on the size and nature of the explosion. Global temperatures fall in the short run because airborne ash shadows the Sun, acting like a giant parasol and cooling the surface. But ash tends to settle back down or is washed out by rain within a few days; its effects can only continue as long as the eruption persists, producing fresh ash. Long-term cooling effects kick in if the eruption is powerful enough, especially in a vertical direction, to shoot large amounts of sulfur dioxide into the stratosphere. Tens of kilometers above the ground, the sulfur dioxide reacts with water vapor to form sulfate aerosols. Because the aerosols are suspended at heights above where raindrops form, there's no way to wash them out. Instead they linger, reflecting solar radiation back into space and cooling the Earth's surface for a year or more after the eruption. The result is what's called a volcanic winter.

The 1815 eruption of Mount Tambora, the 1883 explosion of Krakatoa, and, most recently, the 1991 blow-out of Mount Pinatubo all gave rise to conditions characteristic of volcanic winters. The much larger eruption of Toba, in prehistoric times, caused such a serious volcanic winter that it may have been a factor in pushing the Earth into the last ice age. Whether supervolcanoes can by themselves cause ice ages is questioned by many geologists. But the Toba super-eruption did take place more or less in the run-up to a marked phase of cooling called the Millennial Ice Age, which, as the name suggests, lasted for about a thousand years. The Earth then warmed up again before settling into the depths of the most recent ice age proper.

Our planet has been through multiple cycles of ice ages over the past two and a half million years, reaching a peak, in each case, after 100,000 years or so – a peak during which most of Europe and North America were covered by glaciers. Although super-eruptions may sometimes play a part in the timing of ice ages, the main trigger is thought to be wobbles in the Earth's axis caused mostly by the gravitational influences of the two biggest planets, Jupiter and Saturn, which nudge and tug on the Earth in different ways over periods of thousands of years. The tilt of the axis can vary by a couple of degrees or so, a small amount but enough to change the way sunlight strikes the planet, which, in turn, is enough to explain the advances and retreats of the polar ice sheets.

Right now we're in the midst of an interglacial, or warm spell, called the Holocene, that's been going on for about 11,500 years. Eventually, the Earth will switch back toward conditions that will ultimately plunge us into another ice age – unless some counter-effect stops or slows it. Given that the Earth is currently in the grip of a rapid warming, thanks largely to man-made greenhouse gas emissions, it's hard to forecast how things will play out. Human-caused global warming may delay the start of the next ice age, and give us something more immediate to worry about in the coming decades than the next big freeze-up. Even if the situation weren't made complicated by human activities, there'd be a debate about the timing of the next big cool-down. It used to be thought that the interglacial before this one had lasted about 10,000 years, so that researchers suspected we were due any time to reach the end of our current warm spell. But several years ago, new evidence emerged that the previous warm spell, which occurred 130,000 years ago, may have persisted for 20,000 years. What's more, some scientists have argued that the present warm era has more in common with an earlier interglacial – the one before last – which lasted for 30,000 years, on which basis we may be only about a third the way through this present balmy spell. On the other hand, many experts say that looking simply at orbital cycles suggests the Earth is teetering close to the edge of a new ice age. Only more research on past climates and the way our climate is changing today will give us a better picture of when to expect the next big chill.

THE WORLD IN DEEP FREEZE

As well as normal ice ages, it's become clear that the Earth has suffered a number of cool-downs that are much more extensive and catastrophic in their effects. Called Snowball Earth events, they involve the almost total encasement of the planet in thick snow and ice, right down to the tropical regions. We know this because geologists have found rocks that have been carried and eroded by glaciers near the present-day equator, along with the fossilized remains of organisms that normally live under Arctic conditions.

The last Snowball happened about 700 million years ago and had a spectacular effect on the evolution of life. Before the global cool-down, life-forms were simple, mostly single-celled organisms dwelling in the sea. During the Snowball phase itself, conditions must have been desperately difficult – harsh enough to wipe out the vast bulk of animals and plants today if they'd been alive at the time. The Earth was either completely frozen over, or mostly frozen except for some kind of slush-covered ocean near the equator with seasonally open water. In any event, conditions were desperately tough. Over most of the planet the only liquid water available lay deep below glaciers, or puddled locally around any areas that experienced volcanic activity. Yet when the ice finally retreated and more normal temperatures returned, an extraordinary thing happened. As life emerged from the deep freeze, it flourished, evolved, and diversified to an astonishing, unprecedented degree. In the so-called Cambrian Explosion, a wild, immense variety of multicellular and macroscopic life suddenly appeared – testimony to the truth of the old adage that what doesn't kill you makes you stronger.

There are just a few minor details about this most recent Snowball that we're not clear about: what caused it, what ended it, and, errr… exactly what happened in between. We really are still that ignorant about the processes involved.

But concerning an earlier Snowball Earth event, also referred to as the Huronian glaciation, which occurred about 2.4 to 2.1 billion years ago, there's a strong suspicion of the cause. New types of microbe had emerged at this time, known as cyanobacteria, which had a novel way of doing photosynthesis, and as a by-product gave off oxygen. To us, oxygen is essential. But to many other life-forms around in that ancient era it was toxic. What's more, it destroyed methane in the atmosphere, a gas that had been very plentiful and highly effective as a greenhouse contributor. By killing off many of the microbes that made methane as part of their metabolism, and also by destroying the methane already present, oxygen deprived the Earth of one of the key ways it had been kept so warm. Gradually, the average temperature of the planet fell until finally it entered a Snowball stage.

Most of the kinds of life we see around us today, including large, complex animals and plants, have never had to face a Snowball event. We and our fellow multicellular cousins have endured ice ages, albeit not easily. But it's hard to see how anything much larger than bacteria could survive millions of years during which virtually the entire world was frozen over. Global temperatures would be so low that at the equator it would feel like modern-day Antarctica. The glaciation would be maintained because ice is highly reflective, so that once it forms a thick, almost unbroken coating, it bounces back most of the incoming solar energy into space. A lack of heat-retaining clouds, caused by water vapor freezing out of the atmosphere, would enhance this effect.

Once the Earth was frozen over, there's the question of how it could ever break free of its icy imprisonment. Volcanoes and other geothermal activity taking place beneath the ice of the planet might be one answer. A major gas given off by volcanoes is carbon dioxide, which is a significant greenhouse gas (though not as effective as methane). The amount of carbon dioxide needed to unfreeze the entire Earth has been put at roughly 350 times what is in the atmosphere today. In other words, to melt Snowball Earth, carbon dioxide would have to make up about thirteen percent of the atmosphere, which is a tremendous amount of gas. However, over millions of years, it's possible that enough of it would accumulate beneath the ice and in the atmosphere to finally begin melting ice in the tropics. Once a band of liquid water and land had been freed, it would be darker than the surrounding ice, enabling it to absorb more solar radiation and start a positive feedback loop of warming followed by more melting. Also, once a sliver of ocean had been opened up, surviving organisms would start to multiply, pumping more greenhouse gases into the atmosphere.

Snowball Earth events are rare – a billion years or more may separate them. Under normal circumstances we'd have nothing to fear from them in the foreseeable future. But these are not normal circumstances. Humans are altering the global ecosystem and climate system at an accelerating rate. Most concern today centers on global warming, not freezing. However, there's a possibility that the two may be linked. Global warming may lead to glaciation and cooling in certain areas of the Earth.

Charles Vörösmarty from the University of New Hampshire and his colleagues found that the average annual discharge of freshwater from the six largest Eurasian rivers into the Arctic Ocean has increased seven percent since 1936.[35] The danger is that this could change the historic ocean circulation pattern which conveys warmth to northern latitudes. Too much freshwater leaking from the land into the Arctic Ocean could reduce or shift the patterns of Atlantic deep water formation and stall the ocean currents that help to bring heat to higher latitudes. In particular the heat-carrying Gulf Stream, which brings warmth for much of the UK and Scandinavia, could be blocked and plunge these areas into a deep freeze.

WAYWARD SUN

Nothing's more important to us here on Earth than a huge glowing ball of hot gases, just a short hop of 150 million kilometers away in space. Without the Sun our world would be dark, permanently frozen, and utterly lifeless. We depend on a steady supply of light and heat from our neighborhood star. But the Sun isn't constant in its energy output and, a bit worryingly, we don't properly understand the limits of its variability.

That the Sun is gradually getting brighter isn't in doubt. The total amount of energy it radiates – its luminosity – is increasing. But this general upward trend is fantastically slow, somewhere between six and ten percent every *billion* years. Of more practical concern to us are variations that take place on the scale of a few years, decades, or centuries. These are important because it's been suggested that changes in solar luminosity can significantly affect Earth's climate, and may even be a major factor in climate change.

The Sun's brightness varies slightly between the maximum and minimum of a roughly eleven-year-long cycle called the sunspot cycle. It wasn't until 1843 that this cycle was discovered, by the German astronomer Samuel Heinrich Schwabe, although subsequent studies found historical records of sunspot variability going back to the early seventeenth century. Over the past three centuries, the average length of a cycle has been 10.7 years, with some as short as nine years and others as long as fourteen. Between 1645 and 1715 there were hardly any sunspots at all – a period known as the Maunder Minimum after the astronomer who made a close study of it.

Sunspots, which appear dark because they're a couple of thousand degrees cooler than the surrounding material (though still much hotter than a blast furnace), are sites of intense magnetic activity. When there are more sunspots, the Sun's light-emitting surface, or photosphere, radiates more strongly. Satellite monitoring of solar luminosity since 1980 has shown that the difference in brightness between the lows and highs of solar activity are usually around 0.1 percent. In extreme cases the brightness changes can be as much as 0.3 percent for a week or more when large sunspot groups and faculae (surrounding bright regions) are involved. Recently, various groups of researchers have reported longer cycles of brightness changes with periods of around 87, 210, 2,300, and 6,000 years, although these results remain uncertain and contentious.

It's known that both long- and short-term variations in solar intensity affect global climate – not surprisingly, since the Sun is our main energy source. What we're not sure about is the extent of these changes. Three to four billion years ago the Sun gave off only about seventy percent of the heat and light it does today. If Earth's atmosphere had the same composition then as it does now, there shouldn't have been any liquid water on the surface. But there's overwhelming evidence for plenty of water around at the time, leading to what's been called the faint young sun paradox. A likely solution to the paradox is that our atmosphere used to be much richer in greenhouse gases than it is at present, so that more of the Sun's heat was retained at surface level. Over the next several billion years, as the Sun's energy output increased, the atmospheric composition changed, in large part because of an influx of oxygen beginning around 2.4 billion years ago with the appearance of cyanobacteria (which take in carbon dioxide and give off oxygen as a waste product of their metabolism).

Many researchers see a strong case for a Sun–climate link in the almost-coincident timing of the Maunder Minimum, when the Sun was nearly sunspot-free for about seventy years, and the so-called Little Ice Age in Europe. Although not a true ice age, this was a period of decidedly lower temperatures extending roughly from the middle of the sixteenth century to the middle of the nineteenth.

We're still hazy on the connection between large and small fluctuations in energy output from the Sun and the climate here on Earth. Important data on this subject were to have been gathered by NASA's Glory satellite, launched on March 4, 2011, but unfortunately the rocket's nosecone failed to open properly and the $28 million probe plunged into the Arctic Ocean.

On January 4, 2008, a tiny black speck appeared on the bright face of the Sun. It didn't look like much even through instruments designed for solar observations. But it was important because the magnetic field of the spot pointed in the opposite direction to the fields of all the other spots on the Sun at the time. The solitary oddball spot marked the start of a new solar cycle which over the next eleven years or so will take the Sun from a point of very low activity to one when sunspots and solar flares are at a peak, probably in 2013, and back again. We may be in for a rough ride, as we'll see in the next section. Looking at the bigger picture, though, recent research suggests that the Sun may enter a period of hibernation after the current solar cycle ends and that a repeat of the Maunder Minimum is at least on the cards.

SUN, SUN, SUN, HERE IT COMES

"Just as sure as the Sun will shine every mornin', every time," sings Justin Timberlake. Nothing is more certain in life than that the Sun will rise and shine and set, and do exactly the same thing the next day and the next. The Sun, by a factor of 10,000 the nearest star to us, seems utterly constant and dependable – an endless giver of steady heat and light.

But that's an illusion. Up close, seen through the eyes of spacecraft equipped to watch the Sun in ultraviolet, X-rays, and visible light, the Sun's surface turns out to be a place of turmoil, unpredictability, and unimaginably violent events. Fountains of gleaming plasma arc into space, and a million-mile-per-hour wind of charged particles streams out in all directions into the Solar System. From time to time, a gargantuan flare erupts, releasing a burst of energy that can have serious consequences for communications and power grids here on Earth.

The key to understanding the dangers as well as the benefits of our neighborhood star is to realize that it isn't just a big ball of hot gas. It's a roiling, spinning mass of plasma, made of positively charged atomic nuclei (mainly hydrogen) and negatively charged electrons darting around at high speed. This constantly moving, writhing plasma generates colossal magnetic fields which can break through the surface, giving rise to the various dynamic features seen in the solar atmosphere and the escape of matter and high-energy radiation into space.

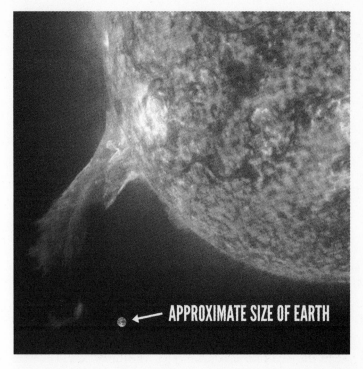

APPROXIMATE SIZE OF EARTH

Figure 14 A giant solar flare erupts from the surface of the Sun. The Earth is shown for comparison. Credit: NASA.

Especially when the Sun is close to a maximum in its cycle, groups of sunspots come together and are sites of intense magnetic activity. It's in the regions around such groups that solar flares and so-called coronal mass ejections originate.

Solar flares are explosions in the solar atmosphere (the outer layers of the Sun) which release huge amounts of energy and radiation across the electromagnetic spectrum from radio waves to gamma rays. The most powerful of them are often associated with coronal mass ejections, which are bursts of charged particles released into space.

When such an ejection is directed towards the Earth, it takes two or three days to arrive and can trigger a geomagnetic storm which causes disturbances in the Earth's magnetic field. Sudden resulting pulses of energy can play havoc with sensitive electrical and electronic gear, both in orbit and on the ground. In 1972, a geomagnetic storm linked to a solar eruption knocked out long-distance phone communications across the state of Illinois. In 1989, another storm plunged six million people into darkness across the Canadian province of Quebec. Today, our dependency on satellite navigation and satellite communications has never been greater, making us all the more vulnerable to unusual activity on the Sun.

Streams of fast-moving protons shot out during coronal mass ejections pose a serious threat to space travelers, especially those on long-duration missions. During the powerful storm of 1989, had there been any astronauts on the Moon they would been exposed to a probably lethal radiation dose of 700 rem. To avoid such exposure on, for example, a journey to Mars will call for some form of shielding, such as a special chamber aboard the spacecraft protected by a strong magnetic or lead lining.

As the next solar maximum approaches in 2013, scientists are keeping a keen watch on the solar weather. On February 15, 2011, the Sun unleashed its strongest flare – a type known as an X-ray flare – in four years. It was picked up by NASA's Solar Dynamics Observatory spacecraft as an intense flash of high-energy ultraviolet light coming from a sunspot. The flare released a coronal mass ejection aimed at the Earth, which caused some communications failures and grounded flights. But, for the most part, our sat-nav systems and power grids escaped more or less unscathed and the only memorable outcome was an unusually bright display of the Northern Lights (aurora borealis).

As 2013 draws near and the Sun builds toward its next extravaganza after one of its quietest periods in fifty years, scientists have warned that we may be hit by a solar storm that's the economic equivalent of a global hurricane Katrina – a technological disaster that could cost up to $2 trillion in damage to communications satellites, power grids, and GPS navigation systems. Society would survive the meltdown but the disruption to commerce, mobile phone networks, and the like would be worse than anything we've previously encountered. Disturbingly, in view of the scale of the threat, the only early-warning system in place is an ageing spacecraft called ACE (Advanced Composition Explorer) launched in 1997. ACE moves in a "halo orbit" around the Sun, located about 1.4 million kilometers away on our sunward side. Any storm blowing in from the Sun has to pass over this spacecraft before it hits Earth. If it took twenty-four hours to go from the Sun to Earth, it would take just under half an hour to reach us from the spacecraft. That brief heads-up is all we'd get before much of our fragile smart technology was overwhelmed.

There may be worse ahead. As we talked about in chapter 4, the day may not be far off when humans begin to connect intimately and more or less permanently, through neural interfaces, with the global web of computers presently called the Internet. If some day we and our machines merge and effectively become a single entity, what would be the harvest if some massive solar flare sent a voltage spike through the circuitry of our planetary brain?

EARTH (4.55 BILLION BCE – 6.75 BILLION CE): R.I.P.

That the Sun will eventually destroy all life on Earth is certain. Over the past four billion years the amount of solar radiation reaching our planet has increased by about a third. Meanwhile the Sun has used up about half the hydrogen fuel in its core, fusing it to form helium, and burns through another four million metric tons of it every second.

The Sun will carry on getting brighter, at the same time growing in size and gradually raising the average temperature of our planet. A billion years from now Earth will have warmed up to the point where the oceans start to evaporate and water vapor in the atmosphere leaks away into space. This process will accelerate until, after another two billion years, the Blue Planet will be blue no more: all the water will have gone, leaving behind a desiccated world on which if life still exists it will be in underground refuges where traces of moisture remain.

Around 5 billion CE, the Sun will run out of hydrogen in its core, although hydrogen-to-helium fusion will continue in a shell around the core. With no radiation pouring out from the center, gravity will squeeze the core, raising the temperature and density until they're high enough to allow helium to start fusing into carbon and oxygen. At the same time as the core is compressed the outer layers of the Sun will swell up, boosting the solar radius by forty percent and doubling the luminosity.

One and a half billion years later, the Sun will be triple the size it is now, with a surface that glows not yellow but orange. Earth will be 100° hotter, and home to nothing except possibly some dregs of extreme heat-loving life-forms.

A quarter of a billion years later still, the Sun will balloon out to become a red giant, about 200 times its present diameter, so that it will engulf Mercury and Venus. Seen from the lava-seas of the then lifeless, molten Earth, it will take up almost half of the sky. Earth may even be entirely consumed or vaporized, returning the material out of which we and it are made to interstellar space. Much depends on the details of solar evolution in these late stages. As a red giant the Sun will lose some of its mass – perhaps up to a third – in the form of a far more vigorous solar wind than what it exhibits today. Despite being bigger, the Sun will have a smaller gravitational pull so that the orbits of the remaining planets, possibly including the Earth, will move outward. Set against this effect will be increased tidal interactions which could tend to make the Earth spiral into the bloated Sun. In any event, it will be academic as far as terrestrial life is concerned. By that time, assuming humans or their successors still exist, they'll presumably have the means to escape destruction, either by relocating within the Solar System or migrating to an entirely new star in a more benign phase of its evolution.

In time, helium fusion will stop inside the solar core but will continue in a shell, itself surrounded by a second hydrogen-burning shell. The Sun will then be even more luminous and unstable, losing matter and varying in brightness. Finally, all nuclear reactions will cease in the Sun, and its outer layers will escape into space to form a so-called planetary nebula, several light years across. Over thousands of years, this nebula will further expand and fade, leaving behind the Sun's dead core, now exposed as a hot, dense ball of degenerate matter smaller than the Earth — a white dwarf.

A REVERSAL IN FORTUNE

It's reassuring to know that whatever the Sun throws at us, we have a strong shield to protect us in the form of our planet's magnetic field. Unfortunately, the geomagnetic field can't always be relied upon.

Every so often, the Earth's magnetic North and South Poles swap places in what's known as a geomagnetic reversal. Evidence that this happens comes from the direction of magnetization in layers of iron-bearing rocks whose ages can be accurately figured out by geologists. The last time the Earth's poles exchanged places was about 780,000 years ago.

Scientists aren't completely clear on why this happens. The

most popular theory is that it has to do with movements in the Earth's liquid iron-nickel core. The magnetic field lines in the core, it's suggested, become tangled and chaotic because of fluid motions and, from time to time, spontaneously flip into the opposite orientation. This is a slow-motion version of what happens to the Sun, whose magnetic field, as we've seen, switches direction every eleven years or so, giving rise to the sunspot cycle and the ups and downs of activity evident on the surface.

During a geomagnetic reversal, the strength of the Earth's magnetic field plummets before recovering again as the new polarity asserts itself, but the details of how much the field strength falls and for how long are sketchy. According to some theoretical models, the flip might take only a year or two; other scenarios suggest a transition period lasting decades. What seems certain is that during the hiatus, the Earth's surface, and any life on it, would be more exposed to the fury of the solar onslaught. For a few years, or maybe a few tens of years, we'd be bombarded by increased levels of high-energy particles normally deflected by our planet's magnetosphere.

It's troubling to learn that we've absolutely no clue when the next reversal will take place because there's no obvious pattern to past reversals. Over the course of the Earth's history the geomagnetic field has flipped back and forth tens of thousands of times, but without any apparent periodicity or regularity. During a long stretch of the Cretaceous period, from 120 million to eighty-three million years ago, the poles stayed put. Afterwards, the frequency of reversals tended to increase, though in no predictable way. During a twelve-million-year span centered about fifteen million years ago, there were no fewer than fifty-one reversals – an average of one every 235,000 years. Given that the last one we know about happened over three-quarters of a million years ago, you might suppose that we're due for another one pretty soon. But, frankly, scientists have no idea. It could start to happen later this century or in another million years' time.

Even if we knew exactly what was going on in the Earth's core, which we certainly don't, we wouldn't be able to forecast the next polar flip more than a decade or two in advance. To test whether we'd be able to spot a flip coming, Gauthier Hulot of Denis Diderot University in Paris, and coworkers, ran computer simulations of the Earth's core based on a range of different physical conditions. They found that no matter what the input, and how precisely the starting conditions were fixed, it was impossible to tell when a polar flip would happen more than a few decades ahead of time. The process appears to be even less predictable than the long-range weather.

Of course, geophysicists nowadays keep a careful watch on what's happening to the Earth's magnetic field and its poles. In recent times, the geomagnetic field has been getting weaker. Its strength has dropped between ten and fifteen percent over the last 150 years, and the rate of decline has sped up in the past few years. Overall, the field strength has plunged by about a third over the past couple of millennia. It's at the state now where it could potentially flip within a few thousand years. But the rate of decrease and the current strength are still within the normal range of variation, as shown by the record of past magnetic fields recorded in rocks, and it's perfectly possible that the field will strengthen again in the future rather than going into reversal.

Unlike the geographical poles (the places on the surface through which the Earth's axis of rotation passes), the magnetic poles wander around. The magnetic north pole, for instance, is presently drifting from northern Canada toward Siberia at an accelerating rate – ten kilometers per year at the beginning of the twentieth century, up to forty kilometers per year in 2003, and since then at an even faster rate. But, again, there's nothing anomalous about this and we can't use it as an argument that a polar reversal is imminent.

All we can be confident about is that another reversal will happen sometime in the future. When it does it could be devastating given our dependence on equipment that would be rendered useless by a sudden major increase in high-energy solar radiation. Technologically, we'd be gravely affected in the interval when the geomagnetic field collapsed and the poles exchanged places. But from a biological perspective it isn't clear that we'd suffer so badly. Our ancestors, such as *Homo erectus*, evidently survived previous reversals, albeit they weren't inconvenienced by breakdowns in sat-nav devices or outages in their power supply. There's also no evidence that polar flips were responsible for past mass extinctions. Life, it seems, can go on despite our planet's occasional magnetic field reversals. One possible reason is that interactions between the solar wind and the Earth's ionosphere may create a sufficiently strong magnetic field to protect the surface from energetic particles even when the internally generated field collapses. Another possibility is that the Earth's field never actually vanishes during a reversal, but is sustained by lots of different poles forming chaotically until the normal two-pole configuration reasserts itself.

SURVIVAL TIPS

A new ice age would put enormous stresses on the human population, forcing people to move away from higher latitudes to where temperatures were more moderate and the ground not permanently frozen. But the good news is that the onset of ice ages is gradual and people would be able to adapt to the changing climate over a period of many decades. This next ice age would also be the first during which humans had advanced technology so that we might be able to counteract or even reverse the effects of glaciation and lower global temperatures, for example by spreading dark material over the polar caps to absorb sunlight and melt the ice, or by slightly altering the composition of the atmosphere to boost its greenhouse effect.

The advent of a volcanic winter, following a major volcanic eruption or super-eruption, would be abrupt and cause much greater problems in the short-term. Agriculture and the food chain in general could be seriously comprised if the eruption was on a very large scale. Where possible people would try to stockpile non-perishable foods or relocate to areas not as badly affected.

CHAPTER 7

TORINO 10

Hurtling toward Earth at twice the speed of the International Space Station is an asteroid bigger than Manhattan Island. Its course is set and unalterable, its target a point in the Atlantic Ocean half an hour's drive west of the coast of Devon in England. Those looking westward across the British Isles and further afield see a rapidly brightening fireball in the sky, rivaling and then far surpassing the brilliance of the Sun. Five seconds before impact the great mountain of rock breaks apart into a single large projectile and several smaller fragments up to hundreds of meters across.

The bulk of the asteroid plunges into the shallow sea and underlying continental shelf unleashing the energy of a billion Hiroshimas. Most of it vaporizes on impact, hurling a vast cloud of sediment and rocky debris, together with water vapor, high into the atmosphere, swiftly blocking out the Sun. Although it's mid-morning in Britain and western Europe, within minutes a grim twilight has given way to total darkness.

Moments after the collision, doomed residents of southwest Britain feel the ground shaking violently as crustal shock waves radiate out from the impact zone. On the heels of this a thermal blast sweeps across the countryside igniting wood and anything else flammable. Soon most of England, including London, is either ablaze or submerged under many feet of salt water. The White Cliffs of Dover have crumbled into the sea.

A wall of water a hundred meters high advances irresistibly in every direction from where the asteroid struck. By late morning much of western Europe is submerged. Northern Germany becomes a series of great lakes, and the flood waters reach as far as Berlin. Racing across the Atlantic, the monstrous tsunami hits the east coast of the United States and Canada. Initially all the water is sucked away from the shore and tankers find themselves stranded in the mud of the ocean floor. Then the monstrous vertical tidal wave, twice as tall as the Statue of Liberty, makes landfall, washing away New York, Boston, and other great seaboard cities in the east, reaching Washington, DC and beyond, leaving the White House accessible only by boat.

By the afternoon of the catastrophe, central Europe is gripped by inky blackness and bitter cold, the only warmth coming from raging forest fires in areas that have escaped the massive flooding. Dry hot winds race over the continent, incinerating most of the vegetation. A global dust cloud blocks sunlight for weeks with only some twilight here or there appearing outside of central Europe.

Where the asteroid struck the crust shatters and crumbles allowing lava to pour out. Close to the impact site were many natural gas reservoirs. These are now ruptured and ignited in a series of enormous explosions. Shock waves from the collision race around and through the planet, triggering volcanic eruptions on the other side of the Earth, in Indonesia and other parts of the Pacific. From these eruptions and explosions, more ash and debris, along with huge amounts of carbon dioxide gas, are pumped into the atmosphere.

Within a week of the impact, many of those who survived the initial blast, fires, and flooding are suffering or dying from asphyxiation as a result of breathing in volcanic dust. Most large mammals are in a similar desperate strait. Oxygen levels plummet as wildfires rage unchecked on every continent. Everywhere there is the stench of decaying corpses.

Plants, too, are dying en masse, starved of light, destroyed in the great fires, or frozen out in the ensuing planet-wide chill. Only fungi and some groups of microbes flourish amid the devastation.

As weeks turn into months, acid rain from sulfur-rich aerosols increasingly pollutes lakes, rivers, and streams, killing off fish and other animals in them. The oceans are large enough to buffer this effect, but smaller basins like the Mediterranean turn into Dead Sea-like environments. The oceanic food chain collapses as plankton are wiped out. Dolphins, whales, and most other large sea dwellers race toward extinction. Sharks and other carrion-eaters thrive for a while on the sudden glut of carcasses, but their fate too is sealed.

On land the vast majority of mammals have disappeared. The survivors are mostly scavengers living off the organic residue of the planet's decimated flora and fauna. Large areas of the Earth are iced over and temperatures are below freezing most of the time, especially at mid-latitudes and further poleward. Only gradually, after several months, does some sunlight begin to break through the dense, debris-laden atmosphere. The first post-apocalypse plants begin to sprout from seedlings that survived the ravages of the impact but these must now contend with soil that has been rendered largely infertile. The total human population has been reduced to a few million. Those who survived, in basements, caves, and underground military installations, now live on stored foods (a dwindling resource), insects, worms, hardy vermin, and some edible fungi which they begin to cultivate.

Ten years later, the massive dust cloud thrown up by the asteroid has largely settled to reveal in most places a barren desert-like landscape. Much of the planet's stock of carbon dioxide has been released into the atmosphere, ramping up the global greenhouse effect so that even in mid-latitudes temperatures often climb as high as 50°C. Tropical and subtropical regions are virtually uninhabitable. Food remains extremely scarce and the remnants of the human race make their way in small groups to subpolar regions. Only slowly does plant cover return to parts of the Earth, and it will take centuries for the scorching temperatures at lower latitudes to abate. Extreme weather conditions will make the survival of any remaining complex life challenging for the foreseeable future.

THE FALLING SKY

Now for the good news: nothing like this is likely to happen anytime soon. We'd almost certainly know about an object this big and potentially destructive that was heading our way over the next few decades. All the same, somewhere out there is a mountain-sized boulder with our name on it. Like all the other planets in the Solar System, our world has been used for target practice by asteroids and comets for billions of years. Over the past ten years or so, thanks to sensitive new instruments and observing programs designed to search for so-called near-Earth objects (NEOs) – asteroids and comets whose orbits can sometimes bring them close to Earth – we're becoming acutely aware just how much stuff there is out there that could be a problem for us unless we set up some form of planetary defense. These new

Figure 15 Artist's impression of an asteroid on a collision course with the Earth. Credit: Jeff Darling.

observational tools are also providing us with an effective early-warning system.

Small asteroids, it turns out, a few meters to a few hundred meters across, whiz past us on a disturbingly regular basis. On March 23, 1989, the 300-meter-wide asteroid Asclepius missed Earth by a mere 700,000 kilometers (just over one and a half times the distance of the Moon) having passed through the exact spot where Earth had been six hours earlier. On March 18, 2004, 2004 FH, an object ten times smaller than Asclepius, slipped by just 43,000 kilometers above the Earth's surface. It had been discovered only three days earlier by the Massachusetts Institute of Technology's Lincoln Near-Earth Asteroid Research (LINEAR) program and would probably have gone unnoticed altogether a decade earlier before such sensitive NEO watch programs were in place. Between 1991 and 2010, ten objects, ranging in size from about five meters to fifty meters across, shot past Earth at less than the distance of the Moon. The largest of these, called 2009 DD45, was about as big as a ten-story building, and came within 63,000 kilometers of Earth (less than one-fifth the distance of the Moon) on March 2, 2004. An object this size coming down in the heart of a major city would be as destructive as a nuclear weapon, flattening buildings across a wide area and killing many thousands of people.

In fact a chunk of space shrapnel roughly the size of 2009 DD45 *did* score a direct hit on our planet back in 1908. Just after 7 AM on June 30 of that year a massive explosion took place near the Podkamennaya Tunguska (Lower Stony) River in Siberia. It was so powerful that it flattened eighty million trees over an area of more than 2,000 square kilometers. A man sitting on the porch of a trading post in Vanavara, sixty kilometers from ground zero, was hurled from his chair and had his shirt set on fire by the heat from the blast. So remote was the area, and so harsh the conditions there during much of the year, that it wasn't until 1927 that a scientific expedition arrived on the scene. What it found was a mystery: no crater, nor any material traces at all of the object that had caused the devastation.

A number of entertaining ideas sprang up over the years to explain the Tunguska Event. Was it perhaps a crashed alien spacecraft? A mini black hole? An antimatter meteorite? In the end, two much more mundane theories survived and astronomers are still divided over which is true. One thing is certain: the body responsible for the Tunguska blast exploded well above the ground, at a height of about 9,000 meters, and packed the punch of about 185 Hiroshima bombs. Either it was a small asteroid or a piece of a comet. In any event, calculations show it must have been a few tens of meters wide and weighed in the region of a hundred thousand tons. Had it blown apart over London or New York, it would have wreaked havoc.

Earth bears the scars of many cosmic impacts. The Barringer Crater, east of Flagstaff, Arizona, is a spectacular formation, over a kilometer wide and 175 meters deep, excavated when a giant iron meteorite, maybe fifty meters across, slammed into the ground about 50,000 years ago. The remains of older, much bigger craters stand out clearly in aerial and satellite photos of Quebec and Saskatchewan provinces in Canada, the West African Shield in Ghana, the Sahara Desert in northern Chad, southern Germany, and elsewhere.

We know, too, of collisions so huge that they changed the world and altered the course of life's evolution. Biggest of all was the run-in that Earth had with a Mars-sized object only 100 million years after our planet formed. That almighty crash hurled a gob of matter, from both bodies, into orbit around the Earth that eventually coalesced to make the Moon. Any signs of life at this early stage would have been erased by the impact. But, with hindsight, Earth's biggest ever collision worked out pretty well for us. Without it there would have been no Moon, and without a big, nearby satellite to help stabilize our planet's axis, Earth would have been subjected to much wider temperature swings, possibly preventing the emergence of large animals. Mars, for instance, has no big moon and has been made to wobble on its axis very significantly over the eons because of the gravitational effects of other planets (most notably Jupiter). Having a large body so close to us means that the total angular momentum of the Earth–Moon system is a lot more than if the Earth were alone, and that means our planet's rotational properties are much harder to alter than if we lacked such a relatively massive companion.

More recently another big blast killed off eighty-five percent of all animal life on Earth, but again had happy consequences if you're a fan of the human race because it allowed the mammals to flourish in the aftermath. It happened about sixty-five million years ago, when a cosmic boulder roughly as large as the one in our opening tale of global mayhem excavated the Chicxulub Crater in what is now Mexico's Yucatan Peninsula. Debate still rages about whether this strike was the sole cause of the mass extinction at the end of the Cretaceous Period, in which the dinosaurs were the most celebrated victims, but it was doubtless a major factor in the wipe-out (perhaps along with the Deccan Traps mentioned in chapter 5). There may have been other asteroid impacts around this time, as hinted at by the discovery of the Boltysh Crater in Ukraine in 2002. Studies of fossil pollen and spores have shown that there was a flourishing of ferns in the wake of this impact. Ferns have an amazing ability to bounce back after environmental catastrophes. A few thousand years later in the fossil record is a second "fern spike" coincident in time with the formation of the Chicxulub Crater. This suggests that the dinosaurs, along with many other animals and plants at the time, were the victims of a double whammy, or perhaps even a whole shower of large space rocks that rained down on the Earth over a period of several millennia. If that proves to be the case, scientists will be left puzzling over what caused the shower. The finger of suspicion will inevitably fall on a collision between two near-Earth objects.

TOO CLOSE FOR COMFORT

NEOs, as their name suggests, are objects that in their travels around the Sun can occasionally come close to the Earth or even, sometimes, ram straight into it. There are three main types: asteroids, meteoroids, and comets. According to the definition used by NEO scientists, near-Earth asteroids (NEAs) range from fifty meters to a few tens of kilometers across and have orbits that never take them more than 1.3 astronomical units from the Sun.

(An astronomical unit is the mean distance of the Earth from the Sun – about 150 million kilometers.) Asteroids are made of rock but often not just a single chunk of rock. Many of the smaller ones seem to be "rubble-pile" objects, composed of rocky debris held together loosely by gravity. A surprisingly high fraction of near-Earth asteroids are binaries (two objects close together, orbiting around each other, some actually in contact). Near-Earth meteoroids are, in NEO terminology, anything smaller than fifty meters across.

Near-Earth comets are a slightly different kettle of fish. Comets are made of more volatile stuff than asteroids; in other words, as well as chunks of rock they contain lots of materials, like water ice and frozen methane, that vaporize if the comet gets into a path that brings it close to the warming rays of the Sun. That's why comets grow tails when they enter the inner parts of the Solar System, and also why comets that repeatedly come close to the Sun eventually burn out so that all that's left behind is a dark porous rocky husk.

With the growing realization that NEOs have caused "extinction-level" events in Earth's history, and that even much smaller ones can have devastating local and regional consequences, scientists, politicians, and the public at large have become concerned with what to do about this danger. The Earth will be hit again, many times, in the future by everything from city-buster Tunguska-type missiles to extinction-class behemoths, unless we do something about the threat.

At the low end of the scale it's been estimated that objects in the five- to ten-meter range enter Earth's atmosphere at the rate of about one a year but most completely disintegrate before they get anywhere near the ground. Tunguska-size intruders, around fifty meters across, are expected on average about once every thousand years. Really big collisions are, thankfully, much rarer. Impacts with NEOs on the order of one kilometer in diameter happen once or twice every million years, while five-kilometer-wide NEOs target us only once every ten million years or so. Still, it's unnerving to think that future encounters are inevitable and could mean the loss of anything from individual cities to our entire civilization.

The thought that we could be snuffed out almost overnight by a big dumb rock has focused attention on what we might do about the risk from wayward asteroids and comets. The first and easiest step is to know the enemy: identify and plot the orbits of every NEO that poses a possible threat to us.

Astronomers made a start on this many years ago as part of a general effort to track down and learn more about asteroids, comets, and other natural debris throughout the Solar System. Since 1947, the Minor Planet Center, now operated as part of the Smithsonian Astrophysical Observatory in Cambridge, Massachusetts, has been collecting data from observers around the world, crunching numbers, and publishing the paths, positions, and other facts and figures to do with every bit of cosmic flotsam known inside the Sun's domain. A growing part of its effort is devoted to NEOs. Every day, torrents of new information course into the Minor Planet Center from amateur astronomers on every continent as well as a number of major professional observation programs, including the University of Arizona's Spacewatch and Catalina Sky Survey, MIT's LINEAR program, and the Siding Spring Survey based at the Siding Spring Observatory in New South Wales, Australia.

HAZARD LEVEL

To date, more than 7,000 near-Earth asteroids have been identified. Between 500 and 1,000 of these are at least one kilometer across, the largest being 1036 Ganymed with a diameter of thirty-two kilometers. In addition, eighty-four near-Earth comets are on astronomers' books. Many of these objects cross the Earth's orbit in their travels around the Sun and so pose a risk to our safety.

To give some measure of the impact hazard of NEOs, a couple of different schemes have been devised. The simplest of these was first put forward by MIT's Richard Binzel in 1995. A revised version of it was adopted by an international group of astronomers at a conference held in Turin (Torino), Italy, in 1999, and subsequently named the Torino scale. It ranks the threat level from 0 (insignificant risk) to 10 (certain collision with global catastrophic consequences). The threat levels are also color-coded, much like those used by some governments to categorize the threat of terrorism, from white, through green, yellow, and orange, to red. The wording that describes each level has been carefully chosen, and was revised in 2005, to avoid any misinterpretations by the public. For example, the description for Level 4 reads: "A close encounter, meriting attention by astronomers. Current calculations give a 1 percent or greater chance of collision capable of regional devastation. Most likely, new telescopic observations will lead to re-assignment to Level 0. Attention by public and by public officials is merited if the encounter is less than a decade away."

Not surprisingly the more sensationalist elements of the media have had a field day overstating the dangers of any newfound NEOs to which astronomers provisionally give a Torino ranking of anything other than 0. That's why the description of Levels 1 through 4 includes the proviso "Most likely, new telescopic observations will lead to a reassignment to Level 0."

Less than one in six NEOs are classed as "potentially hazardous objects", and of these only a handful have ever been honored with a Torino ranking of more than 0. The present record-holder is an asteroid called 99942 Apophis which in 2004 was briefly given a Torino score of 4 because early observations, in the wake of its discovery, suggested there was up to a 2.7 percent probability that it would collide with Earth in 2029. With a diameter of 350 meters and the ability to release around fifty times more energy on impact than the Tunguska Event, Apophis certainly has the clout to kill many millions and devastate several countries if it were to hit our planet in the wrong place, either at sea or on land. But further observations quickly set minds at ease, and any chance of a collision in 2029 was eventually ruled out altogether. There's still a slight chance that during the 2029 fly-by of Earth, Apophis will slip through a few-hundred-meter-wide corridor of space known as a gravitational keyhole, and if it did so it would be nudged onto a collision course for Earth in 2036. Gradually, though, with the benefit of more refined data, even this possibility has become less and less likely.

Thanks to its reputation as the most dangerous space rock known, Apophis has come under keen scrutiny from astronomers over the past few years. It's also been the subject of countless scare stories in the tabloids and on Internet forums. Fortunately, the more we've learned about its orbit the more we've come to realize that Apophis isn't much of a threat to us at all – at least for the foreseeable future. At the time of writing its Torino ranking has been cut to 0 and the odds of an impact in 2036 have lengthened to 1 in 250,000.

It might seem as if we're not in any imminent danger. But the likes of Apophis and the past record of Earth–asteroid collisions, stretching back billions of years, have served as a wake-up call. Scientists and policy-makers have come to realize we need a much clearer picture of what's out there and, eventually, the means to do something about it if an object is found to be on a collision course.

SPACEGUARD

In his 1972 novel *Rendezvous with Rama* Arthur C. Clarke coined the name "Spaceguard" for a fictional program used to give early warning of dangerous near-Earth asteroids. A quarter of a century later, amid mounting evidence that the Earth has suffered catastrophic impacts in the past, the US Congress handed NASA a mandate to find ninety percent of NEAs with a diameter of more than one kilometer – in other words, those that could potentially have global consequences if they struck. This became known as the Spaceguard Goal.

In 2005 US Congressman George E. Brown proposed a bill, which was subsequently passed, to set up an even more comprehensive NEO survey – one that would find, track, and catalogue every near-Earth asteroid and comet bigger than 140 meters across by 2020. A start has been made on that project but there's been a distinct lack of funding to be able to meet the goal. That was the conclusion of a report published in 2010 by the National Research Council called *Defending Planet Earth: Near-Earth Object Surveys and Hazard Mitigation Strategies.* Because Congress never gave NASA any extra money to support its Spaceguard efforts there's no way the space agency can satisfy its mandate over the next decade. The so-called George E. Brown Survey demands more powerful and dedicated observing systems than are currently available.

Yet, despite the financial obstacles, an embryonic form of Spaceguard has been set up. In 1995 the Working Group on Near-Earth Objects of the International Astronomical Union held a workshop entitled *Beginning the Spaceguard Survey* out of which sprang an international organization called the Spaceguard Foundation. This now acts as the coordinating body for various Spaceguard associations that have been set up in countries around the world with the aim of hunting down all NEOs that might deal our planet a serious blow.

One of the latest NEO-tracking scopes to come on-line is the Pan-STARRS PS1 on the summit of Haleakala on the Hawaiian island of Maui. The Panoramic Survey Telescope & Rapid Response System, to give it its full name, uses fairly small mirrors combined with large digital cameras to take detailed pictures of the entire sky several times a month. PS1 is the prototype. Eventually, four identical telescopes will work in harness and enable incredibly detailed surveys of the Solar System and beyond. PS1 alone is expected to discover around 100,000 new asteroids, including many new ones in near-Earth orbits.

DANGER IN THE DARK

The discovery of so many NEOs over the past few years may give the impression that we're getting on top of the problem of the collision threat. To some extent we are, but there continue to be surprises. One of the latest is dark asteroids which were discovered recently by a specially equipped space telescope.

Launched by NASA in December 2009, the Wide-Field Infrared Survey Explorer (WISE) is an orbiting telescope designed to map the entire sky at infrared wavelengths. In its first six weeks of active duty, it uncovered sixteen previously unknown asteroids with orbits close to that of Earth. Nine of these objects reflect less than a tenth of the sunlight that falls on them, making them virtually invisible to telescopes working at optical wavelengths. One of them is as black as fresh asphalt.

Just as disturbing is the fact that these nearby stealth objects have orbits tilted steeply to the plane in which the planets and most other asteroids move around the Sun – the very plane in which the majority of telescopes hunting for asteroids spend most of their time looking. One possibility is that they aren't asteroids at all but former comets which have had their ices vaporized, leaving them tailless and inactive. Many comets have highly inclined orbits and those visited by spacecraft have some of the darkest surfaces ever seen.

Undetected black asteroids or dead comets are just one of the types of NEOs that could spring a nasty surprise on us. Collisions between asteroids can give rise to big chunks of material, some of which could end up targeting the Earth. The fact that more than one large impact seems to have happened around the end of the Cretaceous period could, as we've seen, be explained by a previous smash-up between two near-Earth asteroids. Although rare, such collisions would throw into confusion any previously made calculations about the paths of the two objects concerned, perhaps giving us little warning if one of them then headed directly toward us.

Also hard to predict is the appearance of new comets from deep space. Often, comets never seen before make their way into the inner Solar System from a vast reservoir of frozen comet nuclei called the Oort cloud, which lies far beyond the orbit of Pluto. The Oort cloud is thought to be mainly spherical in shape, extend up to a third of the distance to the nearest star, and contain objects flung out from closer to the Sun by the massive outer planets when the Solar System was very young. New comets swinging into the inner Solar System from the Oort cloud (having had their travel plans altered by faint gravitational nudges, perhaps from passing stars or clouds of interstellar material) can occasionally slam into planets and have the potential to cause massive effects. An instance of this happened back in 1994 when comet Shoemaker-Levy 9 broke apart and then collided with Jupiter, leaving enormous dark scars in the planet's atmosphere which persisted for many months.

Given the enormous number of objects, asteroidal and cometary, out there it's going to be decades before we have a complete catalogue of even the near-Earth variety. And there'll always remain the maverick events, inter-asteroid collisions and new comet appearances, that pop up to surprise us and pose unpredictable, potentially deadly threats to us. Beyond that lies the question of how we respond if an object is on an intercept path.

SURVIVAL TIPS

If something big is coming toward us, we have three basic options: get out of the way, destroy it before it destroys us, or somehow nudge it in time so that it misses us. Future schemes to deal with incoming asteroids and comets fall into these categories – duck, destroy, or deflect.

Of course, we can't move the Earth out of the way. But we could move at least a small fraction of ourselves – the human race – out of immediate harm's way by retreating to underground facilities, equipped with everything we need to survive for months or years until it was safe to venture back above ground. Such facilities already exist, dotted around the world, constructed mainly to protect a privileged few in the event of a nuclear holocaust. But in total they could house only a few million souls while the rest of us were left to fend for ourselves on the surface. Also, if the colliding object were a kilometer or more across, subterranean silos would do nothing to prevent a global catastrophe that could permanently wreck Earth's ecosystem.

Effective methods to prevent impact disasters, locally, regionally, or globally, are of the destroy or deflect variety. Blowing a dangerous space rock apart with nuclear devices might seem the obvious way to eliminate a collision threat. But there are both pros and cons to this brute-force approach. It's the quickest, cheapest, and easiest way to try to mitigate the danger. On the other hand, it might not work – and it might even make the problem worse. Many near-Earth asteroids are just loose aggregates of rock, or "rubble-piles", which would soak up the energy of an explosion in the same way that crumple zones on a car absorb the energy of a collision. Computer simulations have also shown that if an asteroid was broken apart by a nuclear blast, many of the slower-moving bits of the parent body would quickly reassemble. Worst of all, if the fragments remained separate but more or less on the same trajectory as before, the Earth would be peppered by numerous smaller impacts, some of them still radioactive from the nuclear attack, which could cause more damage than one single blow.

Far better would be gradually to shift the orbit of a hazardous object so that it no longer intercepted the Earth. But such a gentle-nudge strategy would take much longer to be effective. One way to move an NEO slowly onto a new course would be by giving it a propulsion system. For example, a series of large rocket engines (fuelled by chemicals or other means) could be landed on and fixed to the surface of the object. How much effect such rockets had would depend on their power, how long they burned for, and the mass of the asteroid. Alternatively, mass drivers might be used. These are electromagnetic catapults, small versions of which have been demonstrated in the laboratory (as early as 1937 by Edwin Northrup), that can fire material into space. Rock mined from the rogue body could be fired repeatedly in small loads to give a small but steady propulsive force. Or a mass driver on the Moon could be aimed at the intruder, spraying it with rock bullets to shift its trajectory.

More subtle deflection ideas have also been suggested. US astronauts Edward Lu and Stanley Love have argued the case for a "gravity tractor" – a very heavy spacecraft hovering above a NEO, with an ion thruster (a much larger version of the kind of engine used to propel spacecraft such as Deep Space 1 and Dawn) to counter the object's gravitational pull, which would gradually tug the larger body away from its original path. This gravitational tug, applied over a long period of time, would be enough to alter the asteroid's heading to miss the Earth narrowly. Another solution put forward is to use unmanned spacecraft to coat one side of an asteroid's surface with a material that would change the rock's reflectivity or thermal conductivity. A large enough change would, over time, modify the asteroid's interaction with sunlight and alter its heading.

No schemes to avoid collisions with NEOs have yet left the drawing board, although some studies may do so over the next few years. The European Space Agency is looking at the design of a space mission, called Don Quijote, that would entail slamming a massive spacecraft or another, smaller near-Earth asteroid, kitted out with rockets, from a high Earth orbit into the target object. In the case of a modest-sized asteroid, like Apophis, calculations have shown that an impacting spacecraft weighing as little as one metric ton could cause a big enough deflection if launched early enough. This is also the preferred strategy of a group of scientists and astronauts known as the B612 Foundation, chaired by Apollo 9 astronaut Rusty Schweickart, which aims to track and predict catastrophic impacts.

Given the range of threats, from known NEOs to undetected dark asteroids and interloping comets, it seems that a variety of approaches may be needed, from fast intercept to decades-long deflection schemes, to defend the planet properly. Until they're in place, we'll have to hope that our luck doesn't run out.

CHAPTER 8

MENACE FROM ACROSS THE LIGHT YEARS

Find the constellation of Orion and you'll soon spot Betelgeuse, the brilliant orange star that marks the eastern shoulder of the Great Hunter. Its color stems from the fact that it's a cool supergiant, one of the largest and brightest stars in our neck of the Galaxy. Put in place of the Sun, Betelgeuse would swallow up the orbits of Mercury, Venus, Earth, and Mars, and reach out almost as far as Jupiter. Although only about ten million years old (compared with the Sun's age of five *billion* years), Betelgeuse is well along its evolutionary track, thanks to its great mass – at least fifteen times that of the Sun.

Massive stars have a lot more nuclear fuel (mainly hydrogen) than their lighter-weight cousins, but they race through this fuel supply many times faster because their cores, where most of the energy production takes place, are so much denser and hotter. When the central hydrogen is gone, the star continues to make light and heat by "burning" (through nuclear fusion) helium, then successively heavier elements, all the way up to iron. But when the stellar core is clogged with iron, the star runs out of options. It takes more energy to fuse iron nuclei than is released in the fusion process.

With the core shut down, there's no longer a strong outward push of radiation to balance the inward force of the star's own gravity. Suddenly, all the material in the star rushes in toward the centre. In the wink of an eye, the core collapses to an incredibly dense state – either a neutron star or a black hole.

The upper layers of the star also fall in, but their implosion quickly turns into an explosion as they're met by a violent surge of energy from the freshly compressed core. The outer layers of the star, including most of the star's mass, are launched into space at speeds of 15,000 kilometers per second or more in a

Figure 16 Artist's impression of a supernova and nearby planet about to be destroyed by the blast. Credit: Jeff Darling.

stupendous blast called a supernova. Along with hot gas, an intense burst of radiation, including ultraviolet, X-rays, and gamma-rays, is released and flies outward at the speed of light. The question is what, if any, effects such a blast could have on us, far across space.

DEATH STARS

At a distance of about 640 light years, Betelgeuse isn't exactly in our stellar backyard but it's among the nearest stars to the Sun that are doomed to go down the supernova route. Exactly when it will explode is anyone's guess. It might already have done so for all we know, given that it takes six centuries for its light to reach us. The best estimate scientists can give is that it will likely blow apart sometime in the next 100,000 years – a mere finger-snap of time by cosmic standards.

When the fateful day does come Betelgeuse will erupt as a so-called Type II supernova. As its outer layers head spaceward at about five percent of the speed of light, its spent core will rapidly shrivel to become, probably, a neutron star some twenty kilometers across. Effectively a solid ball of nuclear matter, a neutron star is so dense that a thimbleful of its contents would outweigh the entire human population.

Seen from Earth, the exploding Betelgeuse will get nearly as bright as the full Moon and be visible for two or three months in broad daylight. But, at our distance, we won't be in any danger because the vast amount of energy released by the supernova will be spread over a bubble of space with a surface area of more than a million square light years.

Type II supernovae don't pose any threat to planets that are hundreds of light years away because their initially deadly radiation spreads out equally in all directions and eventually becomes too thin to be of concern. But there are other kinds of stellar denizens that we do need to fear – in some cases, even if they're pretty remote.

Ageing heavyweight stars, such as Betelgeuse, that manage to hold onto an inflated envelope of gases, right up to the last minute, blow up as Type IIs. Even heavier stars that, toward the end of their lives, shed most or all of their outer layers leaving behind a massive exposed core, explode as Type Ib or Type Ic supernovae, and some of these can be very dangerous indeed.

Deadliest of all are the Type Ic's. These form from the brightest, hottest, most massive stars in existence – stars that are tens of times or even a hundred times heavier than the Sun. In time these stellar monsters develop such a fierce wind of particles, blowing out from their inner regions, that it sweeps the star's outer layers clear away into space. Left behind is a very heavy, hot, exposed core that, at the moment its fusion energy reserves run out, does an extraordinary thing. Most of it is blown to smithereens, but what's left behind quickly collapses to become the densest kind of object in the universe – a black hole.

Black holes themselves are harmless unless you were to get too close to one and fall in, like tumbling over a cliff. The great hazard across interstellar distances comes from the radiation given off during a Type Ic event because it's fired out in a specific direction almost like a laser beam. Along the old star's (or new black hole's) rotation axis a fantastically powerful burst of gamma-rays is shot into space. The burst lasts no longer than the time it takes to read this sentence but in that brief interval unleashes as much energy as the Sun will over its entire lifetime – in fact, more energy in such a short time than can come from any other known process in the Universe. So intense is the pulse that it has the capacity to destroy life on planets that are hundreds or even thousands of light years away. But because the energy from the burst emerges along a narrow jet – anywhere from

2° to 20° wide, observations suggest – its destructive potential is quite tightly focused.

Astronomers think that only a handful of gamma-ray bursts happen in the Milky Way Galaxy every million years, and because of their tight beaming the great majority would be undetectable from our vantage point. The first gamma-ray bursts ever to be seen, located in galaxies far, far away, were spotted accidentally by nuclear watchdog satellites back in the 1960s. A burster within the Milky Way that happens to have its beam shining in our direction is rarer than an honest politician. But they (gamma-ray bursts, that is) must have occurred from time to time during Earth's history and will happen again in the future. A local burster, it's been suggested by a team of researchers at NASA and the University of Kansas, may even have been responsible for one of Earth's five great mass extinctions – the Ordovician-Silurian extinction which happened about 450 million years ago and wiped out over half of all animal life, at a time when most of it still lived in the sea.[36] A repeat of that event today would be catastrophic.

Computer simulations have shown that a ten-second burst, coming from a Type Ic supernova 6,000 light years away, could strip away up to half the Earth's ozone layer. The incoming barrage of gamma-rays would ionize (split apart) nitrogen and oxygen molecules in the atmosphere, forming nitric oxide which acts as a catalyst to destroy ozone. With the planet's radiation shield compromised, a lot more high-energy solar ultraviolet would be able to reach the surface, killing off much of the life on land and within a couple of meters of the top of the ocean. Anything deeper down would survive the radiation exposure but eventually suffer because of the massive disruption to the food chain.

Statistics seem to be on our side, in that gamma-ray bursts possibly "targeted" at us and close enough to be able to cause any harm are thin on the ground. Yet, there are a handful of stars we need to keep an eye on because they're within lethal range and of exactly the type that can become Type Ic supernovae.

One of these potential death stars goes by the unprepossessing name of WR104. "WR" stands for Wolf–Rayet, after two French astronomers, Charles Wolf and Georges Rayet, who first came across this kind of star during their research at the Paris Observatory in 1867. Wolf–Rayets are very hot stars, at least twenty times as massive as the Sun (and often much heavier than that), which are rapidly leaking matter into their surrounding space. The closest example to Earth is one of the stars of the Gamma Velorum system which lies about 800 light years away. WR104 is between six and ten times further away than that in the constellation Sagittarius but has achieved notoriety because optical measurements made with the Keck Telescope in Hawaii over an eight-year period and published in 2008 suggested that its spin axis was pointed more or less directly at us. "Peering down the barrel of a gun" is how one astronomer put it, meaning that if the star did explode it would unleash its deadly bolt of gamma-rays straight at our planet. What's more, we'd have no warning or time to react. Even if we set up sensors at the edge of the Solar System to send a signal to us as soon as they detected an incoming gamma-ray burst, the sensor signal couldn't outrun the gamma pulse which would also be traveling at the speed of light. One moment all would be normal and the next the lethal pulse would be upon us, rapidly dismantling our ozone shield.

Fortunately, other observations of WR104, announced in 2009, paint a slightly less alarming picture. Infrared measurements to do with how hot gas is moving in the vicinity of the Wolf–Rayet star and its close stellar partner – another big, hot, but less evolved star – suggest that the spin axis may actually be aimed a more reassuring 30° or 40° away from our direction. Maybe some alien race will pay the price when WR104 eventually blows itself to bits but the odds have improved that we'll be spared. However, nothing is yet certain. Visually, the pinwheel of dust seen spiraling out from WR104 as a result of colliding stellar winds between the star and its partner does give the distinct impression that we're looking right down the axis of rotation. The infrared observations suggest that this is just an illusion and that we're actually well off-axis. The jury is still out, and only more research will tell us definitely the true state of affairs.

Among other stars in our galactic neighborhood that have the potential to unleash a gamma-ray holocaust is Eta Carinae. This amazing object is roughly at the same distance as WR104 but is even more massive. The heavier of the two super-hot, super-bright stars that make up the Eta Carinae system weighs more than a hundred Suns and is disturbingly unstable. Back in 1843 this stellar disaster-in-the-making became briefly as bright as a supernova without actually exploding. It's one of an elite group of stars that have been seen to undergo what's called a "supernova imposter" event.

Ominously, another star which masqueraded as a supernova in 2004 blew up for real as the supernova called SN2006jc just two years later. Whether Eta Carinae is on the verge of obliteration we'll have to wait and see. It could become a supernova tomorrow, or in tens of thousands of years' time. Luckily for us, its spin axis seems to be pointing elsewhere so that even if it does send out a gamma-ray burst we won't be in the firing line.

The kind of stars that are doomed to explode as Type Ic supernovae aren't exactly shrinking violets. They're the energy tycoons of the Galaxy, easily standing out from the stellar crowd in terms of their sheer brilliance and making them hard to miss. A few more will no doubt turn up in surveys of the Galaxy before we have a complete inventory of them – WR104, for instance, was discovered as recently as 1998 – but it's unlikely we'll find any others so close to Earth that they threaten our survival.

The same can't be said of other potential death stars.

COUNTDOWN TO CATASTROPHE

Gamma-ray bursts, it turns out, come in two distinct types – long and short. Long gamma-ray bursts are the kind we've already talked about. They tend to go on for at least several seconds and, as we've seen, are part of the death throes of very massive stars which blow apart as Type Ic supernovae. Short gamma-ray bursts, on other hand, last under two seconds and seem to have a completely different origin. The most popular theory, at present, is that they're given off when two neutron stars collide. A neutron star, as its name suggests, is made almost entirely of neutrons – the neutral particles found (along with protons) inside atomic nuclei. Because the neutrons are squashed together as tightly as the laws of physics allow, neutron star matter is incredibly dense and neutron stars, despite being up to twice as massive as the Sun, are no more than about twenty-five kilometers across.

More than half of all stars belong to binary or multiple systems in which the components revolve around each other. If two closely orbiting stars both evolve to become neutron stars, they may spiral in towards each other because, in their super-dense state, they steadily lose orbital energy in the form of gravitational waves – ripples in the fabric of space and time, forecast almost a century ago by Albert Einstein. According to Einstein's general theory of relativity, anything with mass causes the surface of space-time to bend, just as a person makes a dip in a trampoline. The bigger and more concentrated the mass, the more severely the local curvature of space-time is altered. If massive objects move around in space-time, the curvature changes to reflect the changed locations of the objects, and if the objects are accelerating, as in the case of orbiting neutron stars, they generate a strong, regular disturbance in space-time which spreads out like ripples in a pond, carrying energy away from the system.

Eventually, as their orbital energy diminishes because of gravitational radiation, the neutron stars get so close together in their frenetic pas de deux that they hurtle around each other a thousand times a second and travel at close to the speed of light. Then they collide, tear each other apart due to mutual tidal forces, and, in less than a hundredth of a second, collapse to form a single black hole. In that action-packed instant, a traffic jam of leftover star matter, still trying to fall into the black hole, gets hopelessly congested and violently explodes, giving rise to a brief but spectacularly powerful gamma-ray burst.

As far as the threat to ourselves goes from a binary neutron star collision, distance and direction of the burst are again key. The problem is that neutron stars can be hard to spot. About 2,000 are known in the whole of our Galaxy, but they're almost certainly the tip of a very large iceberg. We've been able to track down these particular specimens only because they're doing something special that gives away their presence. Many of them are pulsars – their fast rates of spin, powerful magnetic fields, and the chance alignment of their spin axes with our line of sight cause them to beam out radiation towards us as regularly as a lighthouse. Others are members of binary systems with other types of stars and signal their nature by sucking in matter, stolen from their companion, and making it glow fiercely in X-rays and other parts of the spectrum.

Only half a dozen or so binary systems are known in which both members are neutron stars. The first to be found, back in 1974, earned its American co-discoverers, Joseph Taylor and Russell Hulse, the 1993 Nobel Prize in Physics, not so much for the discovery itself but for the indirect proof it enabled of the existence of gravitational waves. After monitoring the system for several years, Taylor and Hulse were able to show that the two objects (one of which is a pulsar) are slowly spiraling inward at exactly the rate expected if they're giving off gravitational waves in accordance with Einstein's general theory of relativity. But although the Hulse-Taylor binary pulsar is ultimately doomed and may well sign off with a brief, galaxy-rivaling gamma-ray burst, the demise isn't expected for another 300 million years or so, and in any case the manic duo is a comforting 21,000 light years away. Likewise, none of the other known binary neutron stars are set to implode for at least tens of millions of years or lie within several thousand light years of Earth.

That isn't to say we're entirely safe from the antics of neutron stars. As mentioned, there are bound to be many binary neutron star systems in the Galaxy lurking unseen. Only if at least one of the stars happens to be a pulsar are they easy to spot. In fact, there might even be a neutron star binary right under our noses – just a few light years away – that's escaped our attention. Also, there are undoubtedly many solitary neutron stars roaming the Galaxy. Two of these solo artists are quite close by, in interstellar terms. The pulsar PSR J0108-1431 is just 424 light years away, while another lone neutron star, called 1RXS J141256.0+792204, could be anywhere from 1,000 to a mere 250 light years away. If all the rogue neutron stars – and solitary black holes, for that matter – were suddenly to be lit up like Christmas tree lights, we'd almost certainly be shocked by their number and, in some cases, their proximity.

Some enterprising science fiction writer could spin an entertaining yarn about two galaxy-wandering neutron stars that just happen to blunder into each other in the solar neighborhood. They collide, decimate each other, and, horror of horrors, in their moment of mutual destruction, send out a short gamma-ray burst aimed straight toward us. In reality, this is about as likely as the annihilation of Earth, as told by Douglas Adams in *The Hitchhiker's Guide to the Galaxy*, in order to make way for an interstellar bypass. Randomly moving stars rarely collide.

SIRIUS CONSEQUENCES

Still on the subject of long shots, how about a nearby star – a *very* nearby star – going supernova and putting an unexpected kibosh on all our plans? A slender 8.6 light years away (just twice the distance of the very nearest star) is the brightest star in the sky, Sirius, a brilliant white jewel of light in the constellation Canis Major, the Great Dog (hence its other name – the Dog Star). It turns out that Sirius isn't a single star but a binary system in which the much smaller and dimmer companion is a white dwarf.

A bit of a mystery surrounds Sirius. According to the folklore of the Dogon tribe from what is now the Republic of Mali in western Africa, Sirius was once red – a claim supported by some sketchy Babylonian, Greek, and Roman references. But today, both stars in the Sirius system are very obviously white. The brighter of the pair, Sirius A, is an "A-type" star, bigger, twice as massive, and much hotter than the Sun, and still making light and heat by fusing hydrogen to helium deep in its interior. Its companion, Sirius B, is also hot, but all of its matter – almost a Sun's worth – is crammed into a ball about the size of the Earth. Now here's the thing: Sirius B used to be the dominant member of this pair. It used to have two and a half times the mass of Sirius A, and was bigger, brighter, and hotter than its mate. Being the heavier of the pair, it evolved faster. Gradually, it swelled to become a red giant and then over time, non-violently, wafted away most of its gassy contents into space. The white dwarf we see now is just the burned-out core of that once great star.

At first sight, the fact that Sirius harbored a red giant in the past seems to tie in with the ancient references to its color. But the timescales don't work. The historical accounts go back at most a few thousand years, whereas astronomers estimate that

Sirius B went through its red giant phase more than 100 *million* years ago. The general feeling among experts is that the old tales of a red Sirius are cases either of misidentity with other objects or of poetic license. On the other hand, there is a way to explain why although Sirius B couldn't have been a red giant in historical times, it may have taken on the appearance of one. White dwarfs come in a variety of compositions. Sirius B belongs to the commonest breed and is made up of a highly compressed carbon-oxygen core surrounded by a thin layer of helium topped off with a skinny "atmosphere" of hydrogen. The suggestion is that a small amount of surface hydrogen might be able to worm its way down into the interior and then, with carbon and oxygen acting as nuclear catalysts, begin fusing into helium. This sudden, brief resumption of energy-making, in an otherwise dead star, would release a pulse of heat which, upon reaching the surface, would cause the hydrogen atmosphere to billow out to thousands of times its normal size. As the atmosphere expanded it would cool and glow bright red. Calculations suggest that after about 250 years the atmosphere would collapse again, losing its ruddy brilliance and returning the white dwarf to its previous state of dim anonymity. Whether this is what actually happened, we simply don't know.

A more spectacular resurgence of the Sirius dwarf could also take place – but not any time soon. Eventually Sirius A will evolve to become a red giant, and like all stars of this type will have only a tenuous hold on the gas in the outer parts of its puffed-up body. Some of this matter will be captured by the dwarf companion, which orbits the larger star at about the distance of Uranus from the Sun. Little by little, Sirius B will gain mass at its partner's expense. At present, B weighs in at 0.98 of a solar mass, making it one of the heaviest white dwarfs known. If it managed to gain another 0.4 of a solar mass by pillaging from its neighbor it would reach a critical mass called the Chandrasekhar limit (after the Indian-born American astrophysicist Subrahmanyan Chandrasekhar who first figured it out).

At the Chandra limit, a white dwarf can no longer support itself against its own inward tug of gravity and collapses to become a neutron star. But if the mass gain is gradual, as it is when a white dwarf in a binary system adds material at its companion's expense, and if also the dwarf is made mostly of carbon and oxygen (which Sirius B is), then something happens before the collapse has chance to take place. According to theory, when a carbon-oxygen dwarf gets within about one percent of the Chandra limit, the pressure and temperature inside it build up to the point at which the carbon and oxygen ignite in a runaway fusion reaction that tears the star apart in an explosion known as a Type Ia supernova.

If Sirius B ever goes supernova in this way, it will suddenly become about five billion times more luminous than the Sun. What's more, although its release of high-energy radiation wouldn't be focused in the manner of a gamma-ray burst, it would happen so close to us, in stellar terms, that the effects would be devastating. We'd be stripped of our life-protecting ozone shield as effectively as if a gamma-ray burst had gone off hundreds of light years away.

But there's no reason to lose any sleep over this. Sirius A isn't scheduled to enter its red giant phase for another billion years or so, and even then the two stars of the Sirius system are probably too far apart for the dwarf to have a sporting chance of looting almost half-a-Sun's worth of extra matter from its partner.

The question is whether there are any other binary systems with white dwarf components, in the Sun's vicinity, which pose a more immediate threat of a Type Ia outburst. Procyon perhaps? It lies only eleven light years away and consists of a star somewhat hotter and brighter than the Sun and a white dwarf companion. But the dwarf, with a lowly starting mass of 0.6 that of the Sun, could never gain enough weight to get into Type Ia territory. We have to go out to a distance of 150 light years and a binary system called IK Pegasi to find the nearest realistic Type Ia supernova candidate – in fact, the nearest supernova candidate to us of any ilk.

The two stars in the IK Pegasi system orbit each other with an average separation of only thirty-one million kilometers, which is less than the distance of Mercury from the Sun. The primary, IK Pegasi A, is bigger and brighter than the Sun but is still a normal, or "main-sequence", star in the sense that it's burning hydrogen to helium in its core. Millions more years will go by before it swells to become a red giant. Its nearby neighbor, B, is unusually heavy for a white dwarf, weighing in at 1.15 solar masses. When A does enter its giant phase, B will be perfectly placed, at such close range, to steal matter from the giant's atmosphere and add to its own mass, steadily pushing it toward the Chandra limit. At some point, there's a good chance it will explode in a Type Ia event.

An important factor to bear in mind when assessing stellar threats is that all stars are moving relative to one another as they orbit around the centre of the Galaxy. As it happens, IK Pegasi is speeding away from us at just over twenty kilometers per second, equivalent to one light year every 14,700 years. This means that if IK Pegasi goes supernova in, say, five million years' time (and it will probably take much longer than that) it would by then be around 500 light years away – too remote to be a safety hazard.

SUPERFLARES

By stellar standards, the Sun is benign and predictable – in fact, an all-round excellent star to have on our doorstep. It's of a breed

Figure 17   Artist's impression of huge flare on the young star EV Lacertae. Credit: NASA/Casey Reed.

called a G-type dwarf: not too hot, nor too cold, yellow in color, long-lived enough that brainy life-forms like us have plenty of time to evolve on a nearby planet.

But not all G-type dwarfs behave alike it seems. Even among its kind, the Sun is of a more placid variety. In 1999, Yale astronomer Bradley Schaefer and his colleagues reported on nine stars, lying within a few tens of light years of Earth, which are Sun-like in most ways but startlingly different in one important respect: they're prone to gargantuan flares that are anywhere from 100 to ten million times more powerful than the largest flare ever detected on the surface of the Sun.[37] Unlike the comparatively mild eruptions on the Sun, these "superflares" affect not just a small patch of the surface but the entire star.

High-energy flares have long been known to be a feature of

infant stars, close-orbiting pairs of stars, and fast-spinning stars. But it came as a shock to find them associated with supposedly normal stars like the Sun. The question has to be asked if it's possible that the Sun itself could put on such a show and, if so, what the outcome would be for us, perched precariously on the third planet.

The nine stars for which Schaefer and his team found (from historical records) evidence of superflares are disturbingly similar to the Sun. One them, Pi[1] Ursae Majoris, is forty-seven light years distant, and a dead ringer for the Sun in size, mass, and brightness. Only in age – a mere 200 million years – is it noticeably different. Another is Kappa[1] Ceti, which lies just thirty light years away and is 800 million years old. So Sun-like is Kappa[1] Ceti that it's considered a prime target for projects searching for terrestrial-type planets. But life on such a world wouldn't be made easy by the star's tendency to blast out dazzling bursts of deadly radiation every century or so.

If the Sun were suddenly to give off a superflare, the least that would happen is that it would fry all satellite electronics, blow out power grids around the world, and disrupt the ionosphere enough to prevent long-distance radio communications. That would be at the very low end of the superflare scale. In the worst case scenario of a powerful superflare, a gaping hole would be blown in the ozone layer and the magnetosphere pushed to ground level on the sunward side, exposing the Earth's surface to X-rays and far ultraviolet radiation for a couple of years and devastating the food chain from the bottom up.

But a big superflare would also have major effects further out in space. It would melt the surfaces of icy moons in the outer Solar System, forming flood plains over the illuminated hemispheres which would later freeze over as smooth as a skating rink. There's some comfort in the knowledge that no evidence of such widespread melting has been seen. There's also nothing in

Earth's recorded history to suggest that we've suffered the effects of a superflare.

One reason may have to do with the way the planets in the Solar System are laid out. According to Eric Rubenstein, a colleague of Schaefer's at Yale, a normally temperate G-type dwarf star can be transformed into a superflaring monster by having a big planet nearby with a strong magnetic field of its own.[38] The planet's field, Rubenstein theorized, becomes more and more tangled up with that of the star, and at some point the twisted fields snap apart, releasing a huge burst of pent-up energy – a superflare. Substitute Jupiter for Mercury in the Solar System and the Sun might periodically send out superflares, rendering the Earth uninhabitable.

If this theory is right it has implications for the prospects of finding life in planetary systems with so-called "hot Jupiters" – giant planets in tiny orbits around their home stars. If superflares are common in such systems then they may make it hard or impossible for life to develop on the surfaces of any other accompanying planets. On the other hand, for all we know, it may be that regular bathings in high-energy particles and radiation drive evolution on at breakneck speed through an increased level of mutations. Alternatively, around stars with hot Jupiters we may find that life tends to evolve and develop underground, in caves, or beneath the surface of oceans.

TOO CLOSE FOR COMFORT

According to an old and now discredited theory, known as the catastrophic hypothesis, the planets and smaller objects of the Solar System were formed when another star passed close by the Sun and tore out a tongue of hot matter which then coalesced into the various objects, including the Earth, which now go around the Sun. Even at the time, astronomers realized that the chances of unrelated stars colliding, or almost colliding, with one another is tiny given the vastness of interstellar space and the almost insignificant size of stars by comparison. That's why supporters of the catastrophic hypothesis were gloomy about the prospects of there being many other planets (and hence life) outside the Solar System.

Eventually the catastrophic hypothesis lost out to the so-called nebular hypothesis – a version of which has morphed into the modern theory of how planets are made – in which planetary systems coalesce from a spinning pancake of material left over from the star formation process.

Under normal circumstances the chances of any one star bumping into another, even over the course of billions of years, are next to nothing. The only time we'd have to worry is if our home star happened to be in a tightly packed group, such as a globular cluster, or the inner, more crowded regions of the Milky Way's core.

Collisions between entire galaxies might also seem to increase the risk of individual star crashes. Speaking of which, there's some unsettling news. Raise your eyes to the night sky in the direction of the constellation Andromeda and you'll notice a faint smudge of light. This is the Andromeda Galaxy, which, seen across a gulf of over two million light years, is the furthest object readily visible without binoculars or a telescope. Like our own Milky Way Galaxy, it's a big spiral system boasting hundreds of billions of stars and a waistline roughly 100,000 light years across. And it's moving quite swiftly our way.

Every second, Andromeda closes the gap between us by 120 kilometers. This line-of-sight, or radial, speed is easy to measure because it causes a so-called blue shift in the light we receive from our galactic neighbor. Much less certain is Andromeda's side-to-side, or transverse, motion. But assuming the worse case scenario of a head-on impact, Andromeda will collide with the Milky Way in about the year 4 billion CE. By that time humans, or their successors, if they exist at all, will likely have evolved into something unimaginable and not be too concerned by intergalactic fender-benders. In any case, the Earth will have become uninhabitable as the ageing Sun grows monstrously to become a red giant.

Still, it's fun to speculate what the consequences might be for life on an Earth-like planet if Andromeda does come charging into us. The answer is: probably not much. Although photos of galaxies make it seem as if they're tightly packed with stars, the fact is that the average distance between stars is mind-bogglingly huge compared with the size of the stars themselves. For example, the nearest star system to the Sun is just over four light years, or about forty trillion kilometers, away. The Sun's diameter, by contrast, is a mere 1.4 million kilometers, so you could fit almost thirty million Suns, side by side, in the yawning gap to the nearest star.

Interstellar space is, for the most part, astonishingly empty. The typical density of stars in a galaxy is about the same as if half a dozen oranges were scattered inside a hollow ball the size of the Earth. If those oranges were moving around at a snail's pace, what are the chances of two of them ever colliding? Virtually zero – and such is the case with stars.

Now imagine an Earth-sized volume of space containing a handful of evenly spaced oranges passing through another Earth-sized volume of space containing a few widely scattered apples. The ridiculously tiny odds of an apple smashing straight into an orange give some idea of the likelihood of stars in our own Galaxy coming head to head with Andromedan interlopers. Of all the hundreds of billions of stars and millions of years involved, maybe six extra collisions, according to one estimate, will result from the galactic incursion.

That isn't to say that there won't be major effects if the Milky Way and Andromeda ever meet. Most likely the galaxies will merge to form, over a period of a billion years or so, a giant elliptical galaxy. There'll also be lively bursts of star formation as interstellar gas and dust in one system ram into their counterparts in the other. But star-on-star impacts will be fantastically rare.

RAIN OF TERROR

An approaching star wouldn't, however, have to run straight into the Sun, or even plough through the heart of the Solar System, to wreak havoc on Earth. All it would have to do is come close enough to disturb the host of icy objects that inhabit the Oort cloud.

No one has ever seen the Oort cloud, but we know it's there because it's the only feasible place from which new comets can arise – a vast reservoir of frozen, cometary nuclei stretching out to roughly a light year from the Sun, or one-quarter of the distance to Proxima Centauri, the nearest star. The Oort cloud is surely home to many billions of chunks of ice and rock ranging in size from about a kilometer to tens or even hundreds of kilometers across. Occasionally some of these objects are deflected so that they plunge toward the inner Solar System, developing tails as they approach the Sun and occasionally becoming bright enough to be seen with the unaided eye.

When comets smash into planets, they can leave a nasty stain or, in the worst case, massive ecological disruption. We just happen to be going through a stage in Earth's history when the risk of cometary collisions isn't very high. But that could all change if another star comes too close and, because of its powerful gravitational influence, like some aggrieved

Olympian deity, hurls a few million Oort cloud objects our way.

Fortunately, most kinds of stars can't creep up on us suddenly, without us noticing them. The only exceptions would be solitary neutron stars or stellar black holes which don't give off much light or other radiation and so could evade detection from a distance. For the most part, though, we have a pretty good handle on the Sun's stellar neighbors and how far away they are and how they're moving.

Using computers to project into the future it's possible to see which of the stars in the solar vicinity are likely to come anywhere near the Sun over the next few million years. The one to watch out for is Gliese 710, a seemingly innocuous orange dwarf star, smaller and cooler than the Sun, which at present lies sixty-three light years away. It happens to be heading more or less straight for us so that about one and a half million years from now its distance will be a mere 1.1 light years – about the same as the edge of the Oort cloud. Not surprisingly it's expected that Gliese 710, modest though it is among stars, will cause mayhem among the Sun's storehouse of would-be comets. The upshot for our children's children ... (several dozen lines of "children's") ... children will be a hair-raising period when comets and asteroids will come plunging in from deep space at an unprecedented rate.

Hopefully by then our descendants will have an impressive planetary defense system in place by means of which they'll be able to spot potential interlopers from the Oort cloud well in advance and then zap them or push them out of the way before they become a health hazard. Forewarned is forearmed in this case, and perhaps that should be the motto in general for dealing with threats of the stellar variety.

SURVIVAL TIPS

Someday perhaps we won't have to worry about stars that explode or pass by us at uncomfortably close range. Undreamed-of technologies, millions of years more advanced than anything we possess today, might enable us instantaneously to shield ourselves from a gamma-ray burst or guard the entire inner Solar System from wayward comets and space rocks. In the shorter term, if we learned of a specific and imminent threat from an unstable star or star system, we could conceivably erect a kind of orbital parasol to shade the Earth from any deadly radiation (though this would also block out any sunlight coming from that direction).

For now, the best we can do is learn more about how stars work, behave, evolve, and move. We can better familiarize ourselves with the physics of supernovae and their precursors, identify any stars in our part of the Galaxy which look liable to explode, and become more expert in predicting how the final stages of their lives will play out. In time we may come to understand the late evolution of massive stars well enough that we can spot subtle signs of an impending supernova in our vicinity, giving us advance warning of a few years to a century or more. Some form of shielding in space might then be set up to deploy automatically at the first sign of trouble and absorb the anticipated gamma-ray flux.

At any rate, it's important to weigh carefully any proposed ways of improving our survival chances. There may even be some commercial opportunities, as suggested by this Internet posting on a Slashdot forum: "For a limited time we are offering heavy gamma-screen lotion. This specially formulated lotion can provide you with protection for up to 12 seconds. We offer full money back after the neutron star event, if you're not completely satisfied."[39]

CHAPTER 9

DARK CONTACT

Aliens that can change shape and blend in among us. Aliens that enslave the human race, lay waste to our defenses with appalling ease, or take over our minds. Aliens whose youngsters find messy, unpleasant ways to emerge from our bodies. All these possibilities and a myriad of others have been explored in science fiction over the past century or so. Occasionally, as in the movies *E.T.* and *Close Encounters of the Third Kind*, the visitors turn out to be benign. But in the vast majority of films and stories about extraterrestrials (*The Thing*, *Alien*, *Independence Day*, etc.) you just know we're in big trouble right from the start.

Is this how it would play out in real life? Earthlings at the mercy of scary, aggressive interstellar intruders, so far ahead of us we might as well be a kindergarten class armed with pop-guns facing the Navy Seals?

We don't know. We don't know anything about life outside

the Earth: how common it is or what it's like. All we do know, for certain, is that life sprang up here under seemingly less than promising conditions, surprisingly fast, and has kept on going for about four billion years, despite occasional major crises. So, we have the impression that living cells are pretty easy to get off the ground and that life, in general, is tough and adaptable.

Astrobiology (or "exobiology", or even "xenobiology", if you prefer) is a new science, still looking for its first piece of authentic subject-matter. It's an attempt to lift biology from the parochial, Earth-centered state that, of necessity, it's been in for centuries, to a universe-wide field of study, like physics.

As recently as the 1960s it was considered a bad career move to go around calling yourself an exobiologist (the name "astrobiology" wasn't yet in vogue). From the viewpoint of mainstream science it was almost as disreputable as being involved in SETI (the Search for Extraterrestrial Intelligence) which, in turn, in the eyes of some scientists, was tantamount to being in league with UFOlogists.

But all that's changed. Astrobiology started to become respectable following the success of the Viking mission to Mars when two identical spacecraft landed on the Red Planet to look for signs of life in the soil. The fact that their results were inconclusive – negative, some would say – didn't matter in the long run. They showed what could be done in terms of biological investigations by robot probes, even tens of million of kilometers from home.

Since the time of Viking, the prospects for extraterrestrial life, both nearby and far away among the stars, have brightened. The basic ingredients for life as we know it – liquid water, organic (carbon-bearing) material, and suitable energy sources – have turned out to be surprisingly common. There's ample evidence that Mars once had plenty of water on its surface. Some of the

Figure 18 Viking 1 on the surface of Mars. Credit: NASA.

big moons in the outer Solar System, such as Jupiter's Europa, appear to have underground oceans, opening up the prospect of entirely new venues for life that we'd never previously considered. Organic molecules have been found in plentiful supply in interstellar space, and in comets and meteorites, suggesting that life may have been given a kick-start in the early days of the Solar System when planets like the Earth were being heavily bombarded with assorted rocky and icy debris. Astronomers have also found hundreds of planets going around other stars, some of them in the so-called habitable zones of their stars in which a world might have permanent liquid water on its surface.

With all this positive news, it seems as if it'll be only a matter of time before alien life of one kind or another is confirmed. That's when the problems might start.

CARGO OF DEATH

As we saw in chapter 3, infections to which our bodies have zero resistance can be lethal. Bacteria or viruses that have mutated so that our immune systems can't attack them, and against which there are no effective medical counter-measures, threaten the health of millions. That's why scientists take very seriously the possibility of contamination by alien microbes.

When the first Apollo astronauts returned from the Moon they had to spend three weeks in strict quarantine to ensure that they hadn't picked up a lunar version of the flu or some other alien pathogen. The rocks they collected were handled in clean rooms behind glass screens by researchers wearing protective gear just in case they contained tiny extraterrestrial stowaways.

No one really expected to find anything alive on the Moon because it's always been dry and airless. But Mars is a very different proposition. In some ways it would be surprising if, at one time or another, there hadn't been life on the Red Planet, given that it was a much warmer, wetter, and altogether friendlier place in the past. It might still harbor microscopic life today, which is why we'll have to be very careful about how we deal with any Martian soil brought back to Earth by future sample-return probes. Even a small colony of visitors from Planet Four which got loose could have an unfortunate effect on the health of Earthlings, human or otherwise.

How dangerous exobiology might be to us would depend on how similar it was to us at the biochemical level. If it used the same basic substances – similar proteins and nucleic acids, for instance – then it might be able to interact with terrestrial life, either in the form of microbes to which we had no resistance (a theme explored in Michael Crichton's book *The Andromeda Strain*) or as toxins that entered the food chain. This might be a particular problem if there happens to be life on Mars, because it could have been transferred there from Earth in the remote past, by meteorites, and therefore be distantly related to us.

We also have to be careful about contaminating Mars with bugs of our own – although perhaps it's already too late for that, despite our best efforts to sterilize spacecraft before launching them toward biologically sensitive worlds. Back in April 1967, the unmanned American spacecraft Surveyor 3 touched down near Oceanus Procellarum, the Moon's Ocean of Storms. Two and a half years later, Apollo 12 astronauts Pete Conrad and Alan Bean, whose lunar module had landed nearby, walked over to Surveyor 3, recovered its camera, and brought it back to Earth. When NASA scientists examined the camera they found that the polyurethane foam insulation covering its circuit boards contained scores of viable specimens of *Streptococcus mitis*, a harmless bacterium commonly found in the human nose, mouth, and throat. Since the camera had been put straight into a sterile bag for its trip back home, there was only one explanation for the microbes: they must have been on the probe since it left Earth. They'd survived not only the outward journey but thirty-one months on an airless, waterless world subjected to huge monthly variations in temperature and bombardment by hard ultraviolet radiation from the Sun. Conrad later commented: "I always thought the most significant thing that we ever found on the whole ... Moon was that little bacteria who came back and living and nobody ever said [expletive] about it."

Since that time, scientists have turned up further amazing examples of survival by bacteria and other small organisms in the face of extraordinarily harsh conditions. There's no doubt that certain kinds of Earth microbes could survive on Mars for many months or even years. Some of them might even be able to reproduce there slowly. In the light of the Surveyor 3 experience, it seems that aboard the various landers and rovers dispatched to the Red Planet over the years we must have inadvertently sent quite a few living representatives from Earth. If there's similar life on Mars it could presumably hitch a ride back here on future sample-return probes or crewed spacecraft, and then there might be an interesting twist on *The War of the Worlds* plot.

In that tale, Earth microbes saved the day by infecting the brainy but malicious Martians. The real Martian invaders might be microscopic, and we mentally superior but unwitting Earthlings their victims.

TARGET EARTH

The other kind of alien encounter we have to worry about is the intelligent variety. It's reasonable to assume that Earth isn't the only planet in the galaxy where simple, single-celled organisms have evolved to become big, complex, and cerebral. High intelligence may not be nearly as common as life itself, but it would be really surprising if, going around the hundreds of billions of other stars in the Milky Way, there weren't some planets inhabited by creatures at least as smart as ourselves.

Scientists have been looking seriously for signs of extraterrestrial intelligence (ETI), on and off, for over fifty years. Just recently, the hunt has begun in earnest for other Earth-like planets – the kind that would most likely be home to beings like us. But we're new kids on the block, technologically speaking. After all, we only invented powered flight just over a century ago. It's a reasonable bet that if ETI exists it will be ahead of us, and most probably way, way ahead of us – perhaps millions of years more advanced.

Pretty soon we'll be able to look for traces of life on any Earth-like worlds we find, and if we detect life-signs we'll be able to look more closely for evidence of alien technology, in the form, for example, of industrial pollutants in the atmosphere of exo-Earths, or large-scale astro-engineering projects. Other races in the Galaxy, if they exist, must have developed those kinds of remote surveillance capabilities a long time ago. All of which suggests that if, as many scientists suspect, other intelligence exists, then someone, somewhere out there, knows about us.

So, what next? If ETI knows we're here and has been monitoring our progress, perhaps it feels we're becoming interesting enough to investigate further or to be worthy of "first contact". Or perhaps it feels we're becoming a danger, not only to ourselves and our planet, but to the Galaxy at large.

The Australian engineer Ronald Bracewell, who taught at Stanford University, wrote about spacecraft designed to make contact with the inhabitants of other stars. A Bracewell probe would carry all the knowledge of its home world and, upon arrival at a target star, send out a beacon to anyone who was listening in an attempt to share its cargo of information.

Alternatively, an alien race might be more interested in watching us than having a meet-and-greet. It might simply want to keep tabs on us by stationing reconnaissance probes or other monitoring gear in the Solar System in handy places. These might include large stable orbits such as the so-called Lagrangian points of the Earth–Moon system. Several searches have already been made for alien hardware in what seem like promising locations, and although nothing has been found so far these are early days. The Solar System is a big place with plenty of room to play hide-and-seek, and, in any case, it may well be that if extraterrestrial spy craft are watching us they're in some kind of stealth mode that renders them invisible to our instruments.

Aliens may even have visited us in person. A lot of dubious archaeological evidence has been put forward arguing the case for extraterrestrial visitation. But there's nothing unscientific about the general idea that crewed scouting missions, or anthropological teams, have come here in the past from distant worlds to study us more closely. It's the sort of thing we might eventually do ourselves, if we can ever figure out how to break the light barrier – using wormholes, perhaps, or some kind of "warp drive" – and cross interstellar distances in a reasonable amount of time. It's also possible that aliens are here on Earth today, perhaps disguised and living among us. The fact that cranks love such ideas and that all kinds of nonsense is written about alien abduction and government cover-ups doesn't make the concept of alien interest or involvement in our affairs any less valid.

INVASION FORCE

One of the great themes of science fiction, from H. G. Wells's *The War of the Worlds* to *Independence Day*, is the invasion of Earth by smart, powerful aliens bent on taking over our planet. It's a great theme because it's terrifying, perfect for dramatic development and acts of heroism, and just about believable. If other brainy beings in space are anything like us they'll have a ruthless, aggressive streak to them which makes them prone to violence and grabbing whatever they need, especially in times of crisis, without too much regard for the consequences.

Simon Conway Morris, an evolutionary biologist at Cambridge University, has argued there's every reason to suppose that even very smart aliens, who are far ahead of us in other ways, will be driven, at root, by the same primitive impulses of self-survival and greed as ourselves. The way evolution by natural selection works seems almost to guarantee that this is true. Should we then be so eager to search for and communicate with other intelligences among the stars?

A number of scientists have said that we shouldn't be

advertising our presence to extraterrestrials by sending welcome messages aboard our star-bound spacecraft, such as Voyagers 1 and 2, and Pioneers 10 and 11. Michael Archer, professor of biology at the University of New South Wales, thinks that the phonograph records attached to each Voyager probe might effectively serve as giant dinner invitations proclaiming "Come to Earth, we have lots of nice, exotic things to eat." This warning brings to mind a classic episode of *The Twilight Zone* in which seemingly friendly aliens arrive on Earth holding copies of a book called "To Serve Man". What could be better than a race of beings who are eager to do anything for us? Unfortunately, it's all a horrible misunderstanding and the publication is in reality a cookbook!

Already there have been quite a number of efforts to send radio messages to other stars in the hopes that someone might be listening. Another mistake, according to the opponents of CETI (Communication with Extraterrestrial Intelligence). But as one of the pioneers in the search for alien intelligence, the American astronomer Frank Drake, has pointed out: "It's too late to worry about giving ourselves away. The deed is done. And repeated daily with every television transmission, every military radar signal, every spacecraft command."

It's naive to suppose that we can keep ourselves secret from the rest of the Galaxy simply by not beaming out messages or fixing calling cards to our spacecraft. As mentioned earlier, we ourselves – mere technological infants – will soon have the means to detect life, and some of the signatures of intelligent life, on planets going around other stars. Extrapolating from our present rate of progress, it isn't fanciful to suppose we'll be able to detect the night glow of alien cities within a few hundred light years a century or so from now. Beings that are just a few thousand years ahead of us will probably have comprehensive catalogues of life-bearing worlds in the Galaxy and will be keeping a close eye on the most interesting of them.

Knowing that we're here, would an alien race attack us? The main motivation would probably be that we have a nice life-friendly planet which would make a good home for some civilization whose own world was becoming uninhabitable. Perhaps really good Earths are hard to come by.

Aliens might invade to get rid of us because we're considered a menace to the stellar neighborhood, or because we, or our planet, have something they want. It's hard to imagine a group of interstellar pirates crossing many light years just to make off with a bunch of iron ore, a tank full of gasoline, or a truckload of laptops, but who knows? In any event, if we *are* invaded from the stars, there isn't going to be much we can do about it. Any creatures with the technology to travel so fast and far, aren't going to be put off by our primitive weaponry.

We'll have to hope that if aliens do eventually land on the White House lawn they'll be less like the murderous monsters from *Alien* or *Predator* and more like cuddly, friendly E.T. But even if the first meeting proves amicable there'll be dangers ahead for us in encountering a race that's far more advanced.

UNDERDOGS

Humans aren't used to playing second fiddle. We pride ourselves on having a superior intellect on a planet that isn't short of species with good brains, like the whales and dolphins, great apes, crows and parrots, and even the octopus.

What's more, all our experience on Earth suggests that there's room for only one top dog in any given niche. The Neanderthals, for example, couldn't compete with the more technologically adroit modern man. Perhaps, galaxy-wide, intelligence whittles down to one super-race. That might be us, but more likely it's some vastly greater intelligence that we haven't yet come across.

Figure 19 Artist's impression of alien spacecraft attacking the Earth. The town being targeted is Gènovés in Spain. Credit: Jeff Darling.

One of the dangers of visiting aliens is that even if they aren't overtly aggressive, bent on destroying or replacing us in classic Hollywood style, we may still ultimately fall victim to them – overwhelmed by their superior technology, just as indigenous races on Earth have succumbed to more "advanced" explorers and colonizers. Even if we aren't wiped out, our culture and everything else distinctive about us might be lost as we're effectively assimilated by our superiors.

Another danger is that we'd simply give up. Why struggle to create, invent, and discover new theories and things when it's all been done before? Even our most deeply held beliefs about the nature of the universe, free will, self-identity, religion, and philosophy could be rendered obsolete overnight, with shocking psychological effects to us as individuals and a species. Perhaps the worst possible fate for mankind would be to become a mere anthropological curiosity, a race seemingly with immense promise that found itself in an evolutionary cul-de-sac and irrelevant.

ALIEN THOUGHTS

A much-discussed theme in CETI is that any sufficiently advanced aliens are likely to be non-biological. We talked, in an earlier chapter, about the likelihood that, in time, man and machine may become permanently fused together in a part-organic, part-technological synergy. The same progression may be a common and natural part of every race's evolution. If so, then we may find that the vast bulk of ETI is to some extent artificial in nature. We'd then have to hope that it wasn't in some way prejudiced against lowly biological forms.

On the other hand, given the pace with which events are unfolding on Earth, it may be that future encounters between intelligences across interstellar distances don't involve groups of individuals. If we all increasingly become part of a collective information processing system, and perhaps a single consciousness, as current trends suggest, communications may ultimately happen between "world-brains". A situation very much like this was portrayed many years ago by Olaf Stapledon in his remarkable visionary novel *Star Maker*. Such an outcome might seem catastrophic for the human race yet be the inevitable way forward for the Universe as a whole: the emergence of a true cosmic mind.

Another far-reaching idea is similar in some ways to one we talked about in chapter 4 – the notion that we might be living in an artificial universe. But whereas earlier we dealt with the strange idea that everything around us might be an incredibly elaborate computer simulation, we now have to face up to the prospect of a universe that is physically real but may have been created by a race of higher beings.

This disturbing possibility stems from some theoretical work by the American physicist Alan Guth. In 1981, Guth was the first to suggest that, in its first few instants, our universe went through a fantastically rapid phase of expansion called inflation. Fluctuations in the primordial sea of space-time from which the universe is believed to have formed were smoothed out during this spectacular blow-out. Today they're visible as subtle variations in the cosmic microwave background – ripples in the cooling glow of the Big Bang.

If the theory of inflation is right, argued Guth and his MIT colleague Edward Farhi in 1987, then there's a chance it could be used to create new universes in the laboratory. The process would involve concentrating enough energy at a single point. This would cause a bubble of space-time to appear, which if large enough, would undergo inflation and become a universe in its own right. Our universe wouldn't be in danger. Instead the child universe would slip through a space-time passageway known as a wormhole and rapidly disconnect completely. Then the child universe would grown on its own, with its own laws of physics, and evolve in ways that we might not even be able to imagine.

And what if this has already happened? Perhaps our universe is the product of alien scientists tinkering in laboratories in some other space and time. That's a disturbing enough idea in itself, especially if you like your gods to be of the less material variety. Worse still is the thought that these alien creators of ours might still be taking an active part in their experiment, and might decide to tinker with it in ways that would erase us from the scheme of things. There aren't any good megacatastrophes, but being wiped out from a universe that had no meaning in the first place, except as perhaps the subject of some higher being's PhD thesis, would surely be the ultimate tragic ending.

SURVIVAL TIPS

As we've seen, some scientists think it's a bad idea that we're drawing attention to ourselves by intentionally beaming out radio signals to the stars or fixing messages to spacecraft which are on escape trajectories from the Solar System. But if a technologically advanced race out there wants to find out about us, it doesn't need to wait for our messages. There are plenty of other clues, which could be picked up by sufficiently powerful instruments, to suggest that Earth is inhabited by a race of industrialized beings. We can't hide from prying alien eyes and, at present, we lack the means to put up much of a fight if an invasion force arrives from somewhere across the Galaxy.

Without some way to conceal our whereabouts with a planetary invisibility shield or weapons that could stop beings almost inevitably our superiors in the science and engineering departments, the best we can do is learn as much as possible about other life in the universe and the universe itself. We can do that by stepping up the search for extrasolar planets, especially those that may be life-bearing, and putting much more effort into the search for interstellar signals that may tell us something of the nature of other intelligence, if it exists.

As far as contamination by more primitive alien life goes, the world's space agencies already have very strict protocols for sterilization of spacecraft designed to land on other worlds or return samples to Earth.

CONCLUSIONS

The biggest threat to life on Earth comes, oddly enough, from the only creatures who spend a lot of time worrying about their well-being and also think of themselves as being the most intelligent species in town. The fact is we're busily engaged in destroying the very support system upon which we and all other animals and plants depend, while simultaneously finding ever more ingenious ways to kill each other. In many ways the universe would be a safer place without us.

One sliver of good news is that the danger of the final few moments of the movie *Dr Strangelove* coming true has receded a bit from its Cold War height. We seem less likely to blow ourselves up in a collective moment of insanity – or "mutually assured destruction" (MAD), to use the official term for this brand of lunacy. On the other hand, we continue to drift with almost carefree abandon toward environmental catastrophe.

Since 1947, the board of directors of the *Bulletin of Atomic Scientists* at the University of Chicago has maintained a symbolic timepiece called the "Doomsday Clock"[40] which gives an estimate of how close we stand to global disaster. Since its inception, shortly after the Second World War, the minute hand on the clock has been repositioned nineteen times. It was at its least scary (seventeen minutes to midnight) in 1991, immediately following the signing of the Strategic Arms Reduction Treaty by the US and USSR. But by 2010 the minute hand was a mere six minutes away from Zero Hour. The reason? While it used to be that the Doomsday Clock reading was dominated by the threat of global nuclear war, now other factors are taken into account which are at least as concerning. Since 2007, according to the *Bulletin of Atomic Scientists*, it has also factored in climate change and "new developments in the life sciences and nanotechnology that could inflict irrevocable harm."

As we've seen in this book, even if we manage to grow up in time to protect and preserve the environment and avoid nuclear holocaust, there are other credible possibilities by which *Homo sapiens* or our whole world, or even the entire universe, could come unstuck in fairly short order. In some cases, the threat comes from technology itself running out of control or having disastrous, unforeseen consequences. In other cases, the danger (such as that of an Earth-bound asteroid) is natural, and advanced technology offers the only means of averting catastrophe.

We'll never be completely safe because the universe is by nature a hostile place, wracked by huge explosions, strafed by deadly radiations, replete with objects hurtling around haphazardly at crazy speeds. That's what makes it so interesting: our fate isn't completely in our hands. But the way we choose to live, and the way we apply the ever-increasingly powerful tools at our disposal, will largely determine if we make it to see a third millennium.

NOTES

1 Eric Drexler, *Engines of Creation: the Coming Era of Nanotechnology* (Anchor Books, 1987).

2 E. Oberdörster, "Manufactured nanomaterials (fullerenes, C_{60}) induce oxidative stress in the brain of largemouth bass", *Environmental Health Perspectives*, 112 (2004), 1058–1062.

3 Report to US EPA in 2002 by researchers from the Center for Biological and Environmental Nanotechnology (CBEN, Rice University, Houston).

4 Vyvyan Howard, Presentation at Nanotox 2004 conference held in Warrington, Cheshire, England.

5 G. Oberdörster, Z. Sharp, V. Atudorei, A. Elder, R. Gelein, W. Kreyling, and C. Cox, "Translocation of inhaled ultrafine particles to the brain", *Inhal Toxicology*, 16, 6–7 (2004), 437–445.

6 C. A. Poland, R. Duffin, I. Kinloch, A. Maynard, W. A. H. Wallace, A. Seaton, V. Stone, S. Brown, et al., "Carbon nanotubes introduced into the abdominal cavity of mice show asbestos-like pathogenicity in a pilot study", *Nature Nanotechnology*, 3, 7 (2008) 423.

7 Y. Shirai, A. J. Osgood, Y. Zhao, K. F. Kelly, and J. M. Tour, "Directional control in thermally driven single-molecule nanocars", *Nano Letters*, 5 (2005), 2330–2334.

8 F. Tipler, "Extraterrestrial beings do not exist", *Quarterly Journal of the Royal Astronomical Society*, 21, 267 (1981).

9 C. Sagan and W. Newman, "The solipsist approach to extraterrestrial intelligence", *Quarterly Journal of the Royal Astronomical Society*, 24, 113 (1983).

10 Leon Lederman, *The God Particle: If the Universe is the Answer, What is the Question?* (Delta, 1994).

11 R. Casadio, S. Fabi, and B. Harms, "Effect of brane thickness on microscopic tidal-charged black holes", *Physical Review D*, 82, 044026 (2010).

12 Arnon Dar, A. De Rújula, and Ulrich Heinz, "Will relativistic heavy-ion colliders destroy our planet?", *Physical Review Letters B*, 470 (1999), 142–148; F. Calogero, "Might a laboratory experiment destroy planet Earth?", *Interdisciplinary Science Reviews*, 25, 3 (2000), 191–202(12).

13 I. Ya. Aref'eva and I. V. Volovich, "Time machine at the LHC", *International Journal of Geometrical Methods in Modern Physics*, 5, 4 (2008), 641–651.

14 Ray Bradbury, "A sound of thunder", *Collier's Magazine*, June 28, 1952.

15 M. S. Morris and K. S. Thorne, "Wormholes in spacetime and their use for interstellar travel: A tool for teaching general relativity", *American Journal of Physics*, 56, 5 (1988), 395–412.

16 Carl Sagan, *Contact* (Simon and Schuster, 1985).

17 H. B. Nielsen and M. Ninomiya, "Test of influence from future in Large Hadron Collider; a proposal", arXiv:0802.2991v2.

18 On-line critiques of Nielsen and Ninomiya's arXiv preprint: Backreaction Blogspot (http://backreaction.blogspot.com/2007/07/whats-new.html); A Quantum Diaries Survivor website (http://dorigo.wordpress.com/2007/07/21/respectable-physicists-gone-crackpotty/).

19 Crave (news website), "Man arrested at Large Hadron Collider claims he's from the future", April 1, 2010 (http://crave.cnet.co.uk/gadgets/man-arrested-at-large-hadron-collider-claims-hes-from-the-future-49305387/).

20 J. P. Blaizot, J. Iliopoulos, J. Madsen, G. G. Ross, P. Sonderegger, and H. J. Specht, "Study of potentially dangerous events during heavy-ion collisions at the LHC", CERN-2003-001 (2003).

21 J. Ellis, G. Giudice, M. L. Mangano, I. Tkachev, and U. Wiedemann, "Review of the safety of LHC collisions", *Journal of Physics G: Nuclear and Particle Physics*, 35 (2008), 115004.

22 L. M. Krauss and J. Dent, "Later time behavior of false vacuum

decay: possible implications for cosmology and metastable inflating states", *Physical Review Letters*, 100 (2008), 171301.

23 B. J. Childers et al., "Necrotizing fasciitis: a fourteen-year retrospective study of 163 consecutive patients", *American Journal of Surgery*, 68 (2002), 109–116.

24 T. R. Walsh, J. Weeks, D. M. Livermore, and M. A. Toleman, "Dissemination of NDM-1 positive bacteria in the New Delhi environment and its implications for human health: an environmental point prevalence study", *The Lancet Infectious Diseases*, 11, 5 (2011), 355–362.

25 C. A. Pope, R. T. Burnett, M. J. Thun, E. E. Calle, D. Krewski, K. Ito, and G. D. Thurston, "Lung cancer, cardiopulmonary mortality, and long-term exposure to fine particulate air pollution", *Journal of the American Medical Association*, 287, 9 (2002), 1132–1141.

26 E. P. Murchison et al., "The Tasmanian Devil transcriptome reveals Schwann cell origins of a clonally transmissible cancer", *Science*, 327 (2010), 84–87.

27 R. Chitale, "Will chromosome Y go bye-bye?" ABCNews/ Health (2009) (http://abcnews.go.com/Health/MensHealthNews/ story?id=8104217).

28 T. Jacks et al., "Tumor spectrum analysis in p53-mutant mice", *Current Biology*, 4 (1994), 1–7.

29 V. Gill, "Naked mole rat's genome 'blueprint' revealed", BBC Nature News (2011) (http://www.bbc.co.uk/nature/14031978).

30 "Teenager moves video icons just by imagination", press release, Washington University in St Louis, October 9, 2006 (http://news. wustl.edu/news/Pages/7800.aspx).

31 N. Bostrum, "Are you living in a computer simulation?", Philosophical Quarterly, 33, 211 (2002), 243–255.

32 M. G. Jackson, R. Carlson, M. D. Kurz, P. D. Kempton, D. Francis, and J. Blusztajn, "Evidence for the survival of the oldest terrestrial mantle reservoir", *Nature*, 466 (2010), 853–856.

33 D. G. van der Meer, W. Spakman, D. van Hinsbergen, M. L. Amaru, and T. H. Torsvik, "Towards absolute plate motions constrained by lower-mantle slab remnants", *Nature Geoscience*, 3, 1 (2010), 36–40.

34 S. Sparks, S. Self, et al., "Super-eruptions: global effects and future threats: report of a Geological Society of London Working Group" (2005) (http://www.geolsoc.org.uk/supereruptions).

35 J. P. Ericson, C. J. Vörösmarty, S. L. Dingman, L. G. Ward, and M. Meybeck, "Effective sea-level rise in deltas: sources of change and human-dimension implications", *Global and Planetary Change*, 50 (2006), 63–82.

36 A. Melott, B. Lieberman, C. Laird, L. Martin, M. Medvedev, B. Thomas, J. Cannizzo, N. Gehrels, and C. Jackma, "Did a gamma-ray burst initiate the late Ordovician mass extinction?", *International Journal of Astrobiology*, 3, 2 (2004), 55–61.

37 B. E. Schaefer, J. R. King, and C. P. Deliyannis, "Superflares on ordinary solar-type stars", *Astrophysical Journal*, 529 (2000), 1026–1030.

38 E. P. Rubenstein and B. E. Schaefer, "Are superflares on solar analogues caused by extrasolar planets?", *Astrophysical Journal*, 529 (2000), 1031–1033.

39 http://science.slashdot.org/story/05/04/11/1731213/Sea-Life-Wiped-Out-by-Neutron-Star-Collision

40 http://www.thebulletin.org/content/doomsday-clock/timeline

FURTHER READING

Aczel, Amir D. *Present at the Creation: the Story of CERN and the Large Hadron Collider* (Crown, 2010).

Agar, Nicholas. *Humanity's End: Why We Should Reject Radical Enhancement* (MIT Press, 2010).

de Boer, Jelle Zeilinga; Sanders, Donald; and Ballard, Robert. *Volcanoes in Human History: the Far-Reaching Effects of Major Eruptions* (Princeton University Press, 2004).

Breining, Greg. *Super Volcano: the Ticking Time Bomb beneath Yellowstone National Park* (Voyageur Press , 2010).

Buchanan, Allen E. *Beyond Humanity?: The Ethics of Biomedical Enhancement* (Oxford University Press, 2011).

DeGrasse Tyson, Neil. *Death by Black Hole: and Other Cosmic Quandaries* (W. W. Norton & Company, 2007).

Felix, Robert W. *Magnetic Reversals and Evolutionary Leaps: the True Origin of Species* (Sugarhouse, 2008).

Gardner, James N. *The Intelligent Universe: AI, ET, and the Emerging Mind of the Cosmos* (New Page Books, 2007).

Garreau, Joel. *Radical Evolution: the Promise and Peril of Enhancing Our Minds, Our Bodies – and What It Means to Be Human* (Broadway, 2006).

Hallam, Tony. *Catastrophes and Lesser Calamities: the Causes of Mass Extinctions* (Oxford University Press, 2005).

Halpern, Paul. *Collider: the Search for the World's Smallest Particles* (Wiley, 2010).

Irwin, L. N. and Schulze-Makuch, D. *Cosmic Biology: How Life Could Evolve on Other Worlds* (Praxis Publishing, 2011).

Kaku, Michio. *Physics of the Future: How Science Will Shape Human Destiny and Our Daily Lives by the Year 2100* (Doubleday, 2011).

Kurzweil, Ray. *The Singularity Is Near: When Humans Transcend Biology* (Penguin, 2006).

Kusky, Timothy. *Asteroids and Meteorites: Catastrophic Collisions with Earth* (Facts on File, 2009).

Lewis, John S. *Comet and Asteroid Impact Hazards on a Populated Earth: Computer Modeling* (Academic Press, 1999).

MacDougall, Doug. *Frozen Earth: the Once and Future Story of Ice Ages* (University of California Press, 2006).

Parker, Bruce. *The Power of the Sea: Tsunamis, Storm Surges, Rogue Waves, and Our Quest to Predict Disasters* (Palgrave Macmillan, 2010).

Plait, Philip. *Death from the Skies!: These Are the Ways the World Will End ...* (Viking Adult, 2008).

Prothero, Donald R. *Catastrophes!: Earthquakes, Tsunamis, Tornadoes, and Other Earth-Shattering Disasters* (Johns Hopkins University Press, 2011).

Ripley, Amanda. *The Unthinkable: Who Survives When Disaster Strikes – and Why* (Three River Press, 2009).

Savino. John. *Supervolcano* (Career Press, 2008).

Schulze-Makuch, Dirk, and Darling, David. *We Are Not Alone: Why We Have Already Discovered Extraterrestrial Life* (Oneworld, 2010).

Smil, Vaclav. *Global Catastrophes and Trends: the Next Fifty Years* (MIT Press, 2008).

Stager, Curt. *Deep Future: the Next 100,000 Years of Life on Earth* (Thomas Dunne Books, 2011).

Wheeler, J. Craig. *Cosmic Catastrophes: Exploding Stars, Black Holes, and Mapping the Universe* (Cambridge University Press, 2007).

Wilson, Edward O. *The Future of Life* (Vintage, 2003).

Zimmer, Carl. *A Planet of Viruses* (University of Chicago Press, 2011).

INDEX

INDEX